Lecture Notes in Computer Science   8427

Commenced Publication in 1973
Founding and Former Series Editors:
Gerhard Goos, Juris Hartmanis, and Jan van Leeuwen

Editorial Board

For further volumes:
http://www.springer.com/series/7407

H. Jaap van den Herik · Hiroyuki Iida
Aske Plaat (Eds.)

Computers and Games

8th International Conference, CG 2013
Yokohama, Japan, August 13–15, 2013
Revised Selected Papers

 Springer

Editors
H. Jaap van den Herik
Aske Plaat
Tilburg University
Tilburg
The Netherlands

Hiroyuki Iida
JAIST
Ishikawa
Japan

ISSN 0302-9743 ISSN 1611-3349 (electronic)
ISBN 978-3-319-09164-8 ISBN 978-3-319-09165-5 (eBook)
DOI 10.1007/978-3-319-09165-5
Springer Cham Heidelberg New York Dordrecht London

Library of Congress Control Number: 2014944064

LNCS Sublibrary: SL1 – Theoretical Computer Science and General Issues

Printed on acid-free paper

Springer is part of Springer Science+Business Media (www.springer.com)

Preface

This book contains the papers of the 8th Computers and Games Conference (CG 2013) held in Yokohama, Japan. The conference took place during August 13 to August 15, 2013 in conjunction with the 17th Computer and Games Tournament and the 20th World Computer-Chess Championship.

The Computers and Games conference series is a major international forum for researchers and developers interested in all aspects of artificial intelligence and computer game playing. The Yokohama conference was definitively characterized by the progress of Monte Carlo Tree Search (MCTS) and the development of new games. Earlier conferences took place in Tsukuba (1998), Hamamatsu (2000), Edmonton (2002), Ramat-Gan (2004), Turin (2006), Beijing (2008), and Kanazawa (2010).

The Program Committee (PC) was pleased to see that so much progress was made in MCTS and that on top of that new games and new techniques were added to the recorded achievements. Each paper was sent to at least three referees. If conflicting views on a paper were reported, the referees themselves arrived at a proper decision. With the help of many referees (see after the preface), the PC accepted 21 papers for presentation at the conference and publication in these proceedings. As usual we informed the authors that they submitted their contribution to a post-conference editing process. The two-step process is meant (i) to give authors the opportunity to include the results of the fruitful discussion after the lecture into their paper, and (ii) to maintain the high quality threshold of the ACG series. The authors enjoyed this procedure.

The above-mentioned set of 21 papers covers a wide range of computer games and many different research topics. We have grouped the topics into five classes. We mention the classes in the order of publication (with the exception of one paper): Monte Carlo Tree Search and its enhancements (7 papers), solving and searching (7 papers), analysis of a game characteristic (5 papers), new approaches (1 paper), and serious games (1 paper).

We hope that the readers will enjoy the research efforts performed by the authors. Below we reproduce brief characterizations of the 21 contributions taken from the text as submitted by the authors. The authors of the first two publications "Dependency-Based Search for Connect6" and "On Semeai Detection in Monte-Carlo Go" received the shared Best Paper Award of the CG2013. As a courtesy to the award winners we start with a paper on Solving and Searching.

"Dependency-based Search for Connect6" is a contribution by I-Chen Wu, Hao-Hua Kang, Hung-Hsuan Lin, Ping-Hung Lin, Ting-Han Wei, Chieh-Min Chang, and Ting-Fu Liao. In 1994, Victor Allis *et al.* proposed dependency-based search (DBS) to solve Go-Moku, a kind of five-in-a-row game. DBS is critical for threat space search (TSS) when there are many independent or nearly independent TSS areas. Similarly, DBS is also important for the game Connect6, a kind of six-in-a-row game with two pieces per move. However, the rule that two pieces are played per move in Connect6 makes DBS rather difficult to apply in Connect6 programs. This paper is the first attempt to apply DBS in Connect6 programs. The targeted program is NCTU6, which won Connect6 tournaments in the Computer Olympiad twice and defeated many professional players in Man-Machine Connect6 championships. The experimental results show that DBS yields a speedup factor of 4.12 on average, and up to 50 for some hard positions.

"On Semeai Detection in Monte-Carlo Go" by Tobias Graf, Lars Schaefers, and Marco Platzner describes the inability of Monte-Carlo Tree Search (MCTS) based Go programs to recognize and adequately handle capturing races, also known as semeai, especially when many of them appear simultaneously. The inability essentially stems from the fact that certain semeai require deep lines of correct tactical play which is not directly related to the exploratory nature of MCTS. In this paper the authors provide a technique for heuristically detecting and analyzing semeai during the search process of a state-of-the-art MCTS implementation. The strength of the approach is evaluated on game positions that are known to be difficult to handle even by the strongest Go programs to date. The results show a clear identification of semeai and thereby advocate the approach as a promising heuristic for the design of future MCTS simulation policies.

"Efficiency of Static Knowledge Bias in Monte-Carlo Tree Search" is a contribution by Kokolo Ikeda and Simon Viennot. Currently, Monte-Carlo methods are the best known algorithms for the game of Go. The Monte-Carlo simulations based on a probability model containing static knowledge of the game have been shown to be more efficient than random simulations. Some programs also use such probability models in the tree search policy to limit the search to a subset of the legal moves or to bias the search. However, this aspect is not so well documented. In the paper, the authors describe more precisely how static knowledge can be used to improve the tree search policy. They experimentally show the efficiency of the proposed method by a large number of games played against open source Go programs.

"Investigating the Limits of Monte-Carlo Tree Search Methods in Computer Go" is authored by Shih-Chieh Huang and Martin Müller. Obviously, Monte-Carlo Tree Search methods have led to huge progress in computer Go. Still, program performance is uneven - most current Go programs are much stronger in some aspects of the game, such as local fighting and positional evaluation, than in other aspects. Well known weaknesses of many programs include (1) the handling of several simultaneous fights, including the 'two safe groups' problem, and (2) the dealing with coexistence in seki. After a brief review of MCTS techniques, three research questions regarding the behavior of MCTS-based Go programs in specific types of Go situations are formulated. Then, an extensive empirical study of 10 leading Go programs investigates their performance of two specifically designed test sets containing 'two safe groups' and seki situations. The results give a good indication of the state-of-the-art in computer Go as of 2012/2013. They show that while a few of the very top programs can apparently solve most of these evaluation problems in their playouts already, these problems are difficult to solve by global search.

"Programing Breakthrough" by Richard Lorentz and Therese Horey analyzes the abstract strategy board game Breakthrough. The game requires a well balanced attention for strategical issues in the early stages but can suddenly and unexpectedly turn tactical. The strategic elements can be extremely subtle and the tactics can be quite deep, involving sequences of 20 or more moves. The authors thoroughly analyze new and existing features of an MCTS-based program to play Breakthrough. They demonstrate that this approach, with proper adjustments, is quite successful. The current version can beat most human players.

"MoHex 2.0: a Pattern-based MCTS Hex Player", is a contribution by Shih-Chieh Huang, Broderick Arneson, Ryan Hayward, Martin Müller, and Jakub Pawlewicz. It deals with the Monte-Carlo tree search revolution which has spread from computer Go to many areas, including computer Hex. MCTS Hex players are now on par with traditional knowledge-based alpha-beta search players. In this paper the reigning Computer Olympiad Hex gold medallist MoHex, an MCTS player, is strengthened by several improvements leading to MoHex2. The first improvement is replacing a hand-crafted MCTS simulation policy by one based on learned local patterns. Two other improvements are the applications of the minorization-maximization algorithm. The resulting pattern weights are used in both the leaf selection ant the simulation phases of MCTS. All these improvements can increase the playing strength considerably, since the resultant MoHex2.0 is about 250 Elo points stronger than MoHex.

"Analyzing Simulations in Monte-Carlo Tree Search for the Game of Go" is a contribution by Sumudu Fernando and Martin Müller. In Monte-Carlo Tree Search, simulations (or playouts) play a crucial role since they replace the evaluation function used in classical game-tree search and guide the development of the game tree. Despite their importance, not too much is known about the details of how they work. This paper starts a more in-depth study of simulations, using the game of Go, and in particular the program Fuego, as an example. Playout policies are investigated in terms of the number of blunders they make, and in terms of how many points they lose over the course of a simulation. The result is a deeper understanding of the different components of the Fuego playout policy. Consequently, 7 suggestions for closer examination are provided. Finally, a list of fundamental questions about simulation policies is given.

"Anomalies of Pure Monte-Carlo Search in Monte Carlo Perfect Games" written by Ingo Althöfer and Wesley Michael Turner is an interesting paper that ends with 6 open problems. So, the reader is encouraged to read this inspiring research paper and try to find an own contribution. A game is called "Monte-Carlo perfect" when in each position pure Monte-Carlo search converges to perfect play as the number of simulations tend toward infinity. The authors exhibit three families of Monte-Carlo perfect single-player and two-player games where this convergence is not monotonic. Moreover, they give a family of MC-perfect games in which MC(1) performs arbitrarily well against MC(1000).

"Developments on Product Propagation" is written by Abdallah Saffidine and Tristan Cazenave. Product Propagation (PP) is an algorithm to backup probabilistic evaluations for abstract two-player games. So far, it was shown that PP could solve Go problems as efficiently as Proof Number Search (PNS). In this paper, the authors exhibit three domains where PP performs better (see the nuances in the paper) than previously known algorithms for solving games. The compared approaches include alpha-beta search, PNS, and Monte Carlo Tree Search. The authors also extend PP to deal with its memory consumption and to improve its solving time.

"Solution Techniques for Quantified Linear Programs and the Links to Gaming" is a contribution by Ulf Lorenz, Thomas Opfer, and Jan Wolf. Quantified linear programs (QLPs) are linear programs (LPs) with variables being either existentially or universally quantified. QLPs are two-person zero-sum games between an existential and a universal player on the one side, and convex multistage decision problems on the other side. Solutions of feasible QLPs are so called winning strategies for the existential player that specify how to react on moves – well-thought fixations of universally quantified variables – of the universal player to be sure to win the game. To find a certain best strategy among different winning strategies, we propose the extension of the QLP decision problem by an objective function. To solve the resulting QLP optimization problem, we exploit the problem's hybrid nature and combine linear programing techniques with solution techniques from game-tree search. As a result, we present an extension of the Nested Benders Decomposition algorithm by the $\alpha\beta$-algorithm and its heuristical move-ordering as used in game-tree search to solve minimax trees. The applicability of our method to both QLPs and models of PSPACE-complete games such as Connect6 is examined in an experimental evaluation.

"Improving Best-reply Search" is a contribution by Markus Esser, Michael Gras, Mark Winands, Maarten Schadd, and Marc Lanctot. Best-Reply Search (BRS) is a new search technique for game-tree search in multi-player games. In BRS, the exponentially many possibilities that can be considered by opponent players is flattened so that only a single move, the best one among all opponents, is chosen. BRS has been shown to outperform the classic search techniques in several domains. However, BRS may consider invalid game states. In this paper, the authors improve the BRS search technique such that it preserves the proper turn order during the search and does not lead to invalid states. The new technique, BRS+, uses the move ordering to select moves at opponent nodes that are not searched. Empirically, they show that BRS+ significantly improves the performance of BRS in Multi-Player Chess, leading to winning 8.3 percent to 11.1 percent more games against the classic techniques maxn and Paranoid, respectively. When BRS+ plays against maxn, Paranoid, and BRS at once, it wins the most games as well.

"Scalable Parallel DFPN Search", written by Jakub Pawlewicz and Ryan Hayward discusses a new shared-memory parallel version of depth-first proof number search. Based on the serial DFPN $1 + \epsilon$ method of Pawlewicz and Lew, SPDFPN searches effectively even as the transposition table becomes almost full, and so can solve large problems. To assign jobs to threads, SPDFPN uses proof and dis-proof numbers. It uses no domain-specific knowledge or heuristics, so it can be used in any domain. The authors tested SPDFPN on problems from the game of Hex. SPDFPN performs well on our scalability test: on a 24-core machine with a task taking 4.2 hours on a single thread, parallel efficiency ranges from about 0.8 on 4 threads to about 0.74 on 16 threads. SPDFPN also performs well on harder problems: it solved all previously intractable 9×9 Hex opening moves, with the hardest opening taking about 111 days. Also, in 63 days, it solved one 10×10 Hex opening move. It is the first time that a computer (or a human) has solved a 10×10 Hex opening move. The current state of the art in Hex solving may fascinate experts and laymen alike.

"A Quantitative Study of 2 × 4 Chinese Dark Chess" is authored by Hung-Jui Chang and Tsan-sheng Hsu. In this paper, the authors study Chinese dark chess (CDC), a popular 2-player imperfect information game, that is a variation of Chinese chess played on a 2 × 4 gameboard. The 2 × 4 version is solved by computing the exact value of each board position for all possible fair piece combinations. The results of the experiments demonstrate that the initial arrangement of the pieces and the place to reveal the first piece are the most important factors to affect the outcome of a game.

"Cylinder-Infinite-Connect-Four except for Widths 2, 6, and 11 is Solved: Draw" is written by Yoshiaki Yamaguchi, Tetsuro Tanaka, and Kazunori Yamaguchi. Cylinder-Infinite-Connect-Four is a variant of the Connect-Four game played on cylindrical boards of differing cyclic widths and infinite height. In this paper, the authors show strategies to avoid losing at Cylinder-Infinite-Connect-Four except for Widths 2, 6, and 11. If both players use the strategies, the game will be drawn. This result can also be used to show that Width-Limited-Infinite-Connect-Four is drawn for any width. Finally, the authors show that Connect-Four of any size with passes allowed is drawn.

"Havannah and TwixT are PSPACE-complete" is authored by Édouard Bonnet, Florian Jamain, and Abdallah Saffidine. Numerous popular abstract strategy games ranging from Hex and Havannah to Lines of Action belong to the class of connection games. Still, very few complexity results on such games have been obtained since Hex was proved pspace-complete in the early 1980s. The authors study the complexity of two connection games among the most widely played ones, i.e., they prove that Havannah and TwixT are pspace-complete. The proof for Havannah involves a reduction from Generalized Geography and is based solely on ring-threats to represent the input graph. The reduction for TwixT builds upon previous work as it is a straightforward encoding of Hex.

"Material Symmetry to Partition Endgame Tables", by Abdallah Saffidine, Nicolas Jouandeau, Cédric Buron, and Tristan Cazenave describes that many games display some kind of material symmetry. That is, some sets of game elements can be exchanged for another set of game elements, so that the resulting position will be equivalent to the original one, no matter how the elements were arranged on the board. Material symmetry is routinely used in card game engines when they normalize their internal representation of the cards. Other games such as Chinese dark chess also feature some form of material symmetry, but it is much less clear what the normal form of a position should be. The authors propose a principled approach to detect material symmetry. Their approach is generic and is based on solving multiple relatively small subgraph isomorphism problems. They show how it can be applied to Chinese dark chess, dominoes, and skat. In the latter case, the mappings obtained are equivalent to the ones resulting from the standard normalization process. In the two former cases, the authors show that the material symmetry allows for impressive savings in memory requirements when building endgame tables. They also show that those savings are relatively independent of the representation of the tables.

"Further Investigations of 3-Member Simple Majority Voting for Chess" is a contribution by Kristian Toby Spoerer, Toshihisa Okaneya, Kokolo Ikeda, and Hiroyuki Iida. The 3-member simple majority voting is investigated for the game of Chess. The programs Stockfish, TogaII and Bobcat are used. Games are played against the strongest member of the group and against the group using simple majority voting. The authors show that the group is stronger than the strongest program. Subsequently, they investigate the following research question: "under what conditions is 3-member simple majority voting stronger than the strongest member?" To answer this question the authors performed experiments on 27 groups. Statistics were gathered on the situations where the group outvoted the group leader. Two conditions were found. First, group members should be almost equal in strength, while still showing a small, but significant strength difference. Second, the denial percentage of the leader's candidate move depends on the strength of the members.

"Comparison Training of Shogi Evaluation Functions with Self-Generated Training Positions and Moves" is written by Akira Ura, Makoto Miwa, Yoshimasa Tsuruoka, and Takashi Chikayama. Automated tuning of parameters in computer game playing is an important technique for building strong computer programs. Comparison training is a supervised learning method for tuning the parameters of an evaluation function. It has proven to be effective in the game of Chess and Shogi. The training method requires a large number of training positions and moves extracted from game records of human experts; however, the number of such game records is limited. In this paper, they propose a practical approach to creating additional training data for comparison training by using the program itself. They investigate three methods for generating additional positions and moves. Then they evaluate them using a Shogi program. Experimental results show that the self-generated training data can improve the playing strength of the program.

"Automatic Generation of Opening Books for Dark Chess" by Bo-Nian Chen and Tsan-sheng Hsu describes that playing the opening game of Chinese dark chess well is a challenge that depends highly on probability. So far there are no known studies or published results for opening game studies. Although automatic open book generation is a common research topic in many games. Some researchers collect master games to build an opening book; others automatically gather computer-played games as their open books. However, in Chinese dark chess, it is still hard to obtain a strong opening book by the above strategies because few master games have been recorded. In this paper, the authors propose a policy-oriented search to build automatically a selective open book that is helpful in practical game playing. The constructed open book provides positive feedback for computer programs that play Chinese dark chess.

"Optimal, Approximately Optimal, and Fair Play of the Fowl Play Card Game" by Todd Neller, Marcin Malec, Clifton Presser, and Forrest Jacobs analyzes optimal play of the jeopardy card game Fowl Play. The paper starts by presenting play that maximizes the expected score per turn, explaining why this differs from optimal play. After describing the equations of optimal play and the techniques used to solve them, the authors present visualizations of such play and compare them to optimal play of a simpler, related dice game Pig. Next, they turn our attention to the use of function approximation in order to demonstrate the feasibility of a good, memory-efficient approximation of optimal play. Finally, they apply the analytical techniques towards the parameterization, tuning, and improved redesign of the game with komi for optimal fairness.

"Resource, Entity, Action: A Generalized Design Pattern for RTS Games" is authored by Mohamed Abbadi, Francesco Di Giacomo, Renzo Orsini, Aske Plaat,

Pieter Spronck, and Giuseppe Maggiore. In many Real-Time Strategy (RTS) games, players develop an army in real time, then attempt to take out one or more opponents. Despite the existence of basic similarities among the many different RTS games, engines of these games are often built ad hoc, and code re-use among different titles is minimal. The authors created a design pattern called "Resource Entity Action" (REA) abstracting the basic interactions that entities have with each other in most RTS games. The paper discusses the REA pattern and its language abstraction. The authors also discuss the implementation in the Casanova game programing language. Their analysis shows that the pattern forms a solid basis for a playable RTS game, and that it achieves considerable gains in terms of lines of code and runtime efficiency. The conclusion is that the REA pattern is a suitable approach to the implementation of many RTS games.

Acknowledgments

This book would not have been produced without the help of many people. In particular, we would like to mention the authors and the referees for their help. Moreover, the organizers of the three events in Yokohama (see the beginning of this preface) contributed substantially by bringing the researchers together. Without much emphasis, we recognize the work by the committees of the CG 2013 as essential for this publication. One exception is made for Joke Hellemons, who is gratefully thanked for all services to our games community. Finally, the editors happily recognize the generous sponsors The Brain and Mind-Sports Foundation, JAIST, Tilburg University, Tilburg center for Cognition and Communication, ICGA, and Digital Games Technology.

April 2014

H. Jaap van den Herik
Hiroyuki Iida
Aske Plaat

Organization

Executive Committee

Editors

H. Jaap van den Herik
Hiroyuki Iida
Aske Plaat

Program Co-chairs

H. Jaap van den Herik
Hiroyuki Iida
Aske Plaat

Organizing Committee

Aske Plaat (Chair)
Hiroyuki Iida
H. Jaap van den Herik

Setsuko Asakura
Johanna Hellemons
Eva Verschoor

List of Sponsors

The Brain and Mind-Sports Foundation
JAIST
Tilburg University
Tilburg center for Cognition and Communication (TiCC)
ICGA
Digital Games Technology

Program Committee

Ingo Althöfer	Tsan-Sheng Hsu	Sancho Salcedo-Sanz
Petr Baudis	Han-Shen Huang	Maarten Schadd
Yngvi Björnsson	Hiroyuki Iida	Richard Segal
Bruno Bouzy	Tomoyuki Kaneko	David Silver
Ivan Bratko	Graham Kendall	Pieter Spronck
Cameron Browne	Akihiro Kishimoto	Nathan Sturtevant
Murray Campbell	Walter Kosters	David Omid Tabibi
Tristan Cazenave	Yoshiyuki Kotani	Tetsuro Tanaka
Bo-Nian Chen	Clyde Kruskal	Gerald Tesauro
Jr-Chang Chen	Richard Lorentz	Olivier Teytaud
Paolo Ciancarini	Ulf Lorenz	Yoshimasa Tsuruoka
Rémi Coulom	Shaul Markovitch	Jos Uiterwijk
Jeroen Donkers	Hitoshi Matsubara	Erik van der Werf
David Fotland	John-Jules Meyer	Peter van Emde Boas
Johannes Fürnkranz	Martin Müller	Jan van Zanten
James Glenn	Todd Neller	Hans Weigand
Reijer Grimbergen	Nathan Netanyahu	Mark Winands
Matej Guid	Pim Nijssen	Thomas Wolf
Dap Hartmann	Jakub Pawlewicz	I-Chen Wu
Tsuyoshi Hashimoto	Jacques Pitrat	Xinhe Xu
Guy M^c C. Haworth	Christian Posthoff	Georgios Yannakakis
Ryan Hayward	Matthias Rauterberg	Shi-Jim Yen
Shun-chin Hsu	Alexander Sadikov	

The Computers and Games Books

The series of Computers and Games (CG) Conferences started in 1998 as a complement to the well-known series of conferences in Advances in Computer Chess (ACC). Since 1998, eight CG conferences have been held. Below we list the conference places and dates together with the Springer publication (LNCS series no.)

Tsukuba, Japan (1998, November)
Proceedings of the 1st Computers and Games Conference (CG98)
Eds H.J. van den Herik and H. Iida
LNCS 1558, 335 pages.

Hamamatsu, Japan (2000, October)
Proceedings of the 2nd Computers and Games Conference (CG2000)
Eds. T.A. Marsland and I. Frank
LNCS 2063, 442 pages.

Edmonton, Canada (2002, July)
Proceedings of the 3th Computers and Games Conference (CG2002)
Eds J. Schaeffer, M. Müller, and Y. Björnsson
LNCS 2883, 431 pages.

Ramat-Gan, Israel (2004, July)
Proceedings of the 4th Computers and Games Conference (CG2004)
Eds. H.J. van den Herik, Y. Björnsson, and N.S. Netanyahu
LNCS 3846, 404 pages.

Turin, Italy (2006, May)
Proceedings of the 5th Computers and Games Conference (CG2006)
Eds. H.J. van den Herik, P. Ciancarini, and H.H.L.M. Donkers
LNCS 4630, 283 pages.

Beijing, China (2008, September)
Proceedings of the 6th Computers and Games Conference (CG2008)
Eds. H.J. van den Herik, X. Xu, Z. Ma, and M.H.M. Winands
LNCS 5131, 275 pages.

Kanazawa, Japan (2010, September)
Proceedings of the 7th Computers and Games Conference (CG2010)
Eds. H.J. van den Herik, H. Iida, and A. Plaat
LNCS 6515, 275 pages.

Yokohama, Japan (2013, August)
Proceedings of the 8th Computers and Games Conference (CG2013)
Eds. H.J. van den Herik, H. Iida, and A. Plaat
LNCS 8427, 260 pages.

The Advances in Computer Chess/Games Books

The series of Advances in Computer Chess (ACC) Conferences started in 1975 as a complement to the World Computer-Chess Championships, for the first time held in Stockholm in 1974. In 1999, the title of the conference changed from ACC into ACG (Advances in Computer Games). Since 1975, thirteen ACC/ACG conferences have been held. Below we list the conference places and dates together with the publication; the Springer publication is supplied with an LNCS series no.

London, England (1975, March)
Proceedings of the 1st Advances in Computer Chess Conference (ACC1)
Ed. M.R.B. Clarke
Edinburgh University Press, 118 pages.

Edinburgh, United Kingdom (1978, April)
Proceedings of the 2nd Advances in Computer Chess Conference (ACC2)
Ed. M.R.B. Clarke
Edinburgh University Press, 142 pages.

London, England (1981, April)
Proceedings of the 3rd Advances in Computer Chess Conference (ACC3)
Ed. M.R.B. Clarke
Pergamon Press, Oxford, UK, 182 pages.

London, England (1984, April)
Proceedings of the 4th Advances in Computer Chess Conference (ACC4)
Ed. D.F. Beal
Pergamon Press, Oxford, UK, 197 pages.

Noordwijkerhout, The Netherlands (1987, April)
Proceedings of the 5th Advances in Computer Chess Conference (ACC5)
Ed. D.F. Beal
North Holland Publishing Comp., Amsterdam, The Netherlands, 321 pages.

London, England (1990, August)
Proceedings of the 6th Advances in Computer Chess Conference (ACC6)
Ed. D.F. Beal
Ellis Horwood, London, UK, 191 pages.

Maastricht, The Netherlands (1993, July)
Proceedings of the 7th Advances in Computer Chess Conference (ACC7)
Eds. H.J. van den Herik, I.S. Herschberg, and J.W.H.M. Uiterwijk Drukkerij Van Spijk B.V. Venlo, The Netherlands, 316 pages.

Maastricht, The Netherlands (1996, June)
Proceedings of the 8th Advances in Computer Chess Conference (ACC8)
Eds. H.J. van den Herik and J.W.H.M. Uiterwijk
Drukkerij Van Spijk B.V. Venlo, The Netherlands, 332 pages.

Paderborn, Germany (1999, June)
Proceedings of the 9th Advances in Computer Games Conference (ACG9)
Eds. H.J. van den Herik and B. Monien
Van Spijk Grafisch Bedrijf Venlo, The Netherlands, 347 pages.

Graz, Austria (2003, November)
Proceedings of the 10th Advances in Computer Games Conference (ACG10)
Eds. H.J. van den Herik, H. Iida, and E.A. Heinz
Kluwer Academic Publishers, Boston/Dordrecht/London, 382 pages.

Taipei, Taiwan (2005, September)
Proceedings of the 11th Advances in Computer Games Conference (ACG11)
Eds. H.J. van den Herik, S-C. Hsu, T-s. Hsu, and H.H.L.M. Donkers
LNCS 4250, 372 pages.

Pamplona, Spain (2009, May)
Proceedings of the 12th Advances in Computer Games Conference (ACG12)
Eds. H.J. van den Herik and P. Spronck
LNCS 6048, 231 pages.

Tilburg, The Netherlands (2011, November)
Proceedings of the 13th Advances in Computer Games Conference (ACG13)
Eds. H.J. van den Herik and A. Plaat
LNCS 7168, 356 pages.

Contents

Dependency-Based Search for Connect6 1
I-Chen Wu, Hao-Hua Kang, Hung-Hsuan Lin, Ping-Hung Lin,
Ting-Han Wei, and Chieh-Min Chang

On Semeai Detection in Monte-Carlo Go 14
Tobias Graf, Lars Schaefers, and Marco Platzner

Efficiency of Static Knowledge Bias in Monte-Carlo Tree Search 26
Kokolo Ikeda and Simon Viennot

Investigating the Limits of Monte-Carlo Tree Search Methods
in Computer Go 39
Shih-Chieh Huang and Martin Müller

Programming Breakthrough 49
Richard Lorentz and Therese Horey

MoHex 2.0: A Pattern-Based MCTS Hex Player 60
Shih-Chieh Huang, Broderick Arneson, Ryan B. Hayward,
Martin Müller, and Jakub Pawlewicz

Analyzing Simulations in Monte-Carlo Tree Search for the Game of Go. ... 72
Sumudu Fernando and Martin Müller

Anomalies of Pure Monte-Carlo Search in Monte-Carlo Perfect Games 84
Ingo Althöfer and Wesley Michael Turner

Developments on Product Propagation 100
Abdallah Saffidine and Tristan Cazenave

Solution Techniques for Quantified Linear Programs and the Links
to Gaming 110
Ulf Lorenz, Thomas Opfer, and Jan Wolf

Improving Best-Reply Search................................... 125
Markus Esser, Michael Gras, Mark H.M. Winands, Maarten P.D. Schadd,
and Marc Lanctot

Scalable Parallel DFPN Search................................. 138
Jakub Pawlewicz and Ryan B. Hayward

A Quantitative Study of 2 × 4 Chinese Dark Chess 151
Hung-Jui Chang and Tsan-sheng Hsu

Cylinder-Infinite-Connect-Four Except for Widths 2, 6,
and 11 Is Solved: Draw.   163
   Yoshiaki Yamaguchi, Tetsuro Tanaka, and Kazunori Yamaguchi

Havannah and TwixT are PSPACE-complete.   175
   Édouard Bonnet, Florian Jamain, and Abdallah Saffidine

Material Symmetry to Partition Endgame Tables   187
   Abdallah Saffidine, Nicolas Jouandeau, Cédric Buron, and Tristan Cazenave

Further Investigations of 3-Member Simple Majority Voting for Chess.   199
   Kristian Toby Spoerer, Toshihisa Okaneya, Kokolo Ikeda, and Hiroyuki Iida

Comparison Training of Shogi Evaluation Functions with Self-Generated
Training Positions and Moves   208
   Akira Ura, Makoto Miwa, Yoshimasa Tsuruoka, and Takashi Chikayama

Automatic Generation of Opening Books for Dark Chess   221
   Bo-Nian Chen and Tsan-sheng Hsu

Optimal, Approximately Optimal, and Fair Play of the Fowl Play
Card Game   233
   Todd W. Neller, Marcin Malec, Clifton G.M. Presser, and Forrest Jacobs

Resource Entity Action: A Generalized Design Pattern for RTS Games   244
   Mohamed Abbadi, Francesco Di Giacomo, Renzo Orsini,
   Aske Plaat, Pieter Spronck, and Giuseppe Maggiore

Author Index   257

Dependency-Based Search for Connect6

I-Chen Wu$^{(\boxtimes)}$, Hao-Hua Kang, Hung-Hsuan Lin, Ping-Hung Lin,
Ting-Han Wei, and Chieh-Min Chang

Department of Computer Science, National Chiao Tung University,
Hsinchu, Taiwan
{icwu, kangbb, stanleylin, bhlin, tinghan,
aup}@java.csie.nctu.edu.tw

Abstract. Allis proposed dependency-based search (DBS) to solve Go-Moku,
a kind of five-in-a-row game. DBS is critical for threat space search (TSS)
when there are many independent or nearly independent TSS areas. Similarly,
DBS is also important for the game Connect6, a kind of six-in-a-row game with
two pieces per move. Unfortunately, the rule that two pieces are played per
move in Connect6 makes DBS extremely difficult to apply to Connect6 programs. This paper is the first attempt to apply DBS to Connect6 programs. The
targeted program is NCTU6, which won Connect6 tournaments in the Computer Olympiad twice and defeated many professional players in Man-Machine
Connect6 championships. The experimental results show that DBS yields a
speedup factor of 4.12 on average, and up to 50 for some hard positions.

Keywords: Connect6 · NCTU6 · Dependency-based search · Threat-space
search

1 Introduction

Dependency-based search (*DB Search*), proposed by Victor Allis *et al.* [1, 2], is a
search method which explores search under dependency constraints. It was successfully used to solve games such as Double-Letter Puzzle, Connect-four and Go-Moku
[3]. Go-Moku is a kind of five-in-a-row game without prohibited moves. A generalized family of *k*-in-a-row games [9, 10] were introduced by Wu *et al.*, and Connect6 is
an interesting one in the family with the following rules [14]. Two players, named
Black and *White*, alternately place two pieces on empty *squares*[1] of a 19 × 19 Go
board in each turn, except that Black, who plays first, places one piece initially. The
player who gets *k* consecutive pieces of his own first wins. The game is a tie when the
board is filled up without either player winning.

Threats are the key to winning these types of games. For example, in Go-Moku, a
four (a threat) forces the opponent to defend, or the opponent loses. *Threat space
search* (*TSS*) is an important winning strategy for these games, where a winning path
is found through threats. For example, in the Go-Moku community [2], *victory-by-
continuous-fours* (*VCF*) is a well-known strategy, whereby a winning sequence is

---

[1] Practically, stones are placed on empty intersections of Renju or Go boards. In this paper, when we
say squares, we mean intersections.

H.J. van den Herik et al. (Eds.): CG 2013, LNCS 8427, pp. 1–13, 2014.
DOI: 10.1007/978-3-319-09165-5_1, © Springer International Publishing Switzerland 2014

achieved through continuously playing fours (threats). Since the opponent's choice of possible replies is limited, the search space based on threats is greatly reduced. Thus, TSS can search much deeper than regular search methods.

While TSS shows its promise in finding the winning sequence, further improvements are possible when there are independent or nearly independent TSS areas. An example was given in [2] (cf. Diagram 3a of [2]). Hence, Allis *et al.* [11, 12] proposed DBS to help minimize the state space of TSS. The method removes unnecessary *independent threats* and focuses on *dependency relations*. However, some threats may be independent at a given point in time but mutually dependent later on. Allis *et al.* also proposed a method called "combination" to help solve this problem.

The above issue is also critical to Connect6 as well as other *k*-in-a-row games. For Connect6, similarly, there may be independent or nearly independent TSS areas, especially during the end game phase where there are usually many independent TSS areas. Unfortunately, independent TSS areas are more likely to become mutually dependent for Connect6 due to the property that two pieces are played for every move.

This paper is the first attempt to apply DBS to Connect6 programs. We give our definitions and notation for DBS in Connect6 in Sect. 2, propose to construct a dependency-based hypergraph to identify dependencies in Sect. 3, and propose some methods for DBS in Sect. 4. Our experiments were run based on our Connect6 program, named NCTU6, which won Connect6 tournaments [5, 10, 13, 15–17] several times from 2006 to 2011, and defeated many top-level human Connect6 players [6] in man-machine Connect6 championships from 2008 to 2011. Our experimental results in Sect. 5 show that NCTU6 with these methods was speeded up about 4.12 times in average when compared to the original NCTU6. Also, for some hard positions, the speedup was up to 50 times. Section 6 makes concluding remarks. Due to paper size limitation, the background for Connect6 is omitted. We directly follow the terminologies in [9, 10], such as threats, single-threat, double-threat, victory by continuous double-threat-or-more moves (VCDT), and victory by continuous single-threat-or-more moves (VCST).

2 Definitions and Notation

This section gives definitions and notations related to TSS. For simplicity of discussion, let the attacker indicate the player who we want to prove to win, and the defender the other. Let a move $m = (s, s')$ be defined as a move from a position s to another position s'. The set of pieces of move m is denoted by $P(m)$, and the set of threats generated by move m is denoted by $T(m)$. A sequence of moves ψ is

$$\psi = \langle m_1, m_2, \ldots, m_{2t} \rangle, \text{ where } t \in \mathbb{N}.$$

In ψ, all m_{2i+1} moves are attacker moves, while all m_{2i} moves are defender moves. If all attacker moves are double-threat-or-more, the sequence is called a *continuous double-threat-or-more sequence* or a *CDT*. If the last attacker move m_{2t-1} is three-threat-or-more, the attacker wins in this CDT sequence, which is called a *lead-to-win CDT*. If all attacker moves are one-threat-or-more in a sequence, it is a *CST*. In this paper, only CDT is discussed for simplicity.

For example, a sequence $\langle m_a, m_b, m_c, m_d, m_e, m_f, m_g, m_h \rangle$ is illustrated in Fig. 1(a) below. $P(ma)$ contains the two black pieces marked with a. Additionally, $T(ma)$ contains a double-threat, four black pieces highlighted with a blue rectangle in Fig. 1(a). The move m_c contains two single threats (STs) to form a double-threat (DT), so $T(m_c)$ contains eight black pieces. The sequence is a CDT since all Black moves are double-threats, and is a lead-to-win CDT since the last Black move m_g has three threats.

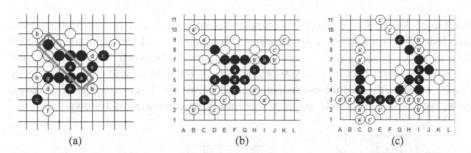

Fig. 1. (a) A lead-to-win CDT, (b) a CCDT with the same attacker moves as (a), and (c) another CCDT.

A lead-to-win CDT ψ does not imply its initial position, denoted by $r(\psi)$, is winning, since a variety of defensive moves are possible for the defender. In order to prove that a given initial position r is winning, we need to prove the following: $\exists m_1, \forall m_2, \exists m_3 \ldots, \forall m_{2t}$ such that a sequence $\langle m_1, m_2, m_3, \ldots, m_{2t} \rangle$ starting from r are CDTs. In the paper [17, 18], it is clear to see that the complexity becomes dramatically smaller if we conceptually allow the defender to play all defensive positions at once, which we refer to as a *conservative defense*. For example, all the White moves in Fig. 1(b) are conservative defenses. Thus, we define the sequence of moves with conservative defenses in Definition 1.

Definition 1: A CDT $\psi = \langle m_1, m_2, \ldots, m_{2t} \rangle$ is called a *conservative CDT* or *CCDT*, if all defender moves m_{2i} are conservative defenses. Since conservative defenses are fixed and unique, we can combine both moves $\langle m_{2i-1}, m_{2i} \rangle$ into a macro-move M_i. Let M_i^A denote attacker move m_{2i-1}, M_i^D defender move m_{2i}. Then, a CCDT can be represented as a sequence of macro moves Ψ as follows.

$$\Psi = \langle M_1, M_2, \ldots, M_t \rangle \text{ where } t \in \mathbb{N}. \qquad \square$$

Since conservative defenses contain all possible defender moves, one important property is: if a CCDT sequence Ψ is lead-to-win, the initial position $r(\Psi)$ is a win for the attacker, and Ψ is also a *VCDT*. As illustrated in Fig. 1(b), $\Psi = \langle M_a, M_b, M_c, M_d \rangle$ is a VCDT, where $M_x = \langle m_x, m_{x'} \rangle$ for all $x = a, b, c, d$.

Definition 2: In a CCDT Ψ defined as above, a macro-move M_j is said to *depend on* M_i, denoted by $M_i \prec M_j$, if $P(M_i^A) \cap T(M_j^A) \neq \emptyset$, where $i < j$. $\qquad \square$

Let us illustrate a dependency relation using the CCDT in Fig. 1(b). We have $M_a \prec M_b$ due to the square at F6. Similarly, we have $M_b \prec M_c, M_a \prec M_d$ and $M_c \prec M_d$, but not $M_b \prec M_d$, denoted by $M_b \nprec M_d$. Another illustration is in Fig. 1(c), where $M_a \prec M_d$ and $M_c \prec M_d$, but $M_a \nprec M_b$ and $M_a \nprec M_c$, implying M_b and M_c are independent of M_a. Thus, M_d is a *combination move* from M_a and M_c.

3 Dependency-Based Hypergraph

For Go-Moku or Renju [8], the dependencies of moves can be represented by a directed graph straightforwardly. However, for Connect6 (where each move includes two pieces), if we directly draw the dependencies between macro-moves, the dependency graph becomes very sophisticated. In order to simplify the dependency graph, we split those double-threat moves with two single-threats into *half-moves*. Now, we classify all double-threat moves into three categories as follows.

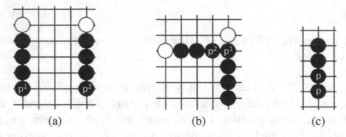

Fig. 2. Three categories of double-threat moves. (a) T^I_{ST} (b) T^D_{ST} (c) T_{DT}.

- T^I_{ST}: the set of double-threat moves M with two independent single-threats. Each of the two attacker pieces, p^1 and p^2 in $P(M^A)$, generates one single-threat independently. More specifically, let the macro-move be divided into two *half-moves*, $M^1 = M^{1A}, M^{1D}$ and $M^2 = \langle M^{2A}, M^{2D} \rangle$, where $M^{1A} = \{p^1\}$, $M^{2A} = \{p^2\}$, and M^{1D} and M^{2D} are the sets of conservative defenses to M^{1A} and M^{2A} respectively. An example is shown in Fig. 2(a). Obviously, there is no dependency between M^1 and M^2. Both M^1 and M^2 are *primitive macro-moves* or *primitive moves*, defined as the moves with the least number of pieces which can generate threats.
- T^D_{ST}: the set of double-threat moves M with two dependent single-threats. Let p^1 and p^2 denote the two attacker pieces in $P(M^A)$. Without loss of generality, let M^1 be a half-move containing p^1 and generating one single-threat, and let M^2 be another containing p^2 and generating another single-threat depending on M^1. An example is shown in Fig. 2(b). According to the definition of dependency, we have $M^1 \prec M^2$. Both M^1 and M^2 are also primitive moves.
- T_{DT}: the set of other double-threat moves, which are usually generated from live twos. An example is shown in Fig. 2(c). In this case, the macro-move cannot be divided into two half-moves since each piece alone would not form a threat. The move M itself is a primitive move.

For a VCDT sequence $\Psi = \langle M_1, M_2, \ldots, M_t \rangle$, let all macro-moves be translated into primitive moves as above. Then, the dependencies of these primitive moves can be drawn into a directed acyclic hypergraph, called a *dependency-based hypergraph* (*DBH*) in this paper. Formally, a DBH $\mathbb{G}$ is defined to be $(\mathbb{V}, \mathbb{E})$, where $\mathbb{V}$ is the set of vertices and $\mathbb{E}$ is the set of hyperedges. Each vertex V in $\mathbb{V}$ represents a position. The root, V_r, represents the initial position and has an indegree of 0. All other vertices have an indegree of 1; that is, for every V other than V_r, there is a single corresponding incoming hyperedge E.

Each hyperedge $E = (\{V_1, V_2, \ldots, V_k\}, V)$ in $\mathbb{E}$ represents a primitive move, where $\{V_1, V_2, \ldots, V_k\}$ are the sources and V is the destination. Let E_i be the corresponding hyperedge of V_i. Then, the condition $E_i \prec E$ must hold. On the other hand, for every primitive move E' where $E' \prec E$, one of the hyperedges $E_1, E_2, \ldots, E_k$ must represent E'. However, if E does not depend on any moves, then $E = (\{V_r\}, V)$. For consistency, let E^A denote the attacker move of E and E^D the defender move.

```
Procedure BUILDPSI(Ψ)
  1:   for every macro-move  M ∈ Ψ do
  2:       ADDMOVE(M)
  3:   end for
end
```

```
Procedure ADDMOVE(M)
  1:   if M ∈ T_DT then
  2:       create hyperedge  E  for the DT;
  3:       ADDE(E);
  4:   else
  5:       create hyperedges  E¹ and  E²  for each ST;
  6:       // without loss of generality, let  E¹ ≺ E²  if  M ∈ T_ST^D.
  7:       ADDE( E¹);
  8:       ADDE( E²);
  9:   end if
end
```

```
Procedure ADDE(E)
  1:   Let  V_source = ∅  be the set of source vertices of  E;
  2:   for every active hyperedge  E' ∉ E do
  3:       if  P(E'^A) ∩ T(E^A) ≠ ∅  then
  4:           add the destination vertex of  E'  to  V_source;
  5:       end if
  6:   end for
  7:   if  V_source = ∅  then set  V_source = {V_r};
  8:   if  E ∉ E  then add  E  into  E;
  9:   mark  E  as active;
end
```

The DBH $\mathbb{G}$ is constructed from macro moves one by one. For simplicity, we first consider building $\mathbb{G}$ from moves in a given CCDT $\Psi = \langle M_1, M_2, \ldots, M_t \rangle$ by the procedure BUILDPSI shown as above. In practice, during the search, moves are added to $\mathbb{G}$ as they are traversed, so BUILDPSI is not actually used. Each move M is checked in ADDMOVE and the corresponding hyperedge and vertex are generated. In the case that $M \in \mathcal{T}_{DT}$, M is transferred into one hyperedge E and inserted into $\mathbb{G}$ via ADDE. In the case that $M \in \mathcal{T}_{ST}^I \cup \mathcal{T}_{ST}^D$, M is split into two half-moves and then respectively translated into two hyperedges E^1 and E^2, which are then inserted into $\mathbb{G}$ (let $E^1 \prec E^2$ if $M \in \mathcal{T}_{ST}^D$) using ADDE. In the rest of this paper, $\mathbb{E}(M)$ denotes the set of hyperedges corresponding to M, namely $\{E^1, E^2\}$ for both $\mathcal{T}_{ST}^I$ and $\mathcal{T}_{ST}^D$, and $\{E\}$ for $\mathcal{T}_{DT}$.

In the procedure ADDE for adding a hyperedge E, find all the hyperedges E' in $\mathbb{G}$ that satisfy $E' \prec E$, then let the destination of E' be one of the sources of E. If no dependent hyperedges are found, put V_r into the set of sources of E.

Furthermore, DBH can be constructed incrementally during threat-space search. Assume we have searched up to M_t in a CCDT $\Psi = \langle M_1, M_2, \ldots, M_t \rangle$. The hyperedges (in $\mathbb{E}(M_t)$) added via ADDE are marked as active in line 9 of ADDE. When the search on M_t is finished and the search recurses to M_{t-1}, we simply mark as inactive the hyperedges that were marked as active for M_t. The active hyperedges form a subgraph of $\mathbb{G}$, denoted by $\mathbb{G}(\Psi)$, which consists of all the hyperedges corresponding to primitive moves of $M_1, M_2, \ldots, M_t$. Note that the inactive hyperedges are still in $\mathbb{G}$. For illustration, the DBH that is constructed for the CCDT in Fig. 1(b) is shown in Fig. 3(a), and another DBH resulting from Fig. 1(c) is in Fig. 3(b).

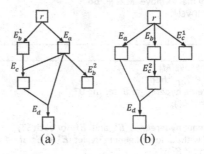

Fig. 3. (a) A DBH constructed for the CCDT in Fig. 1(b), (b) another for the CCDT in Fig. 1(c).

Definition 3: E' is said to be *reachable* from E if there exists some path from E to E', denoted as $E \ll E'$. Similarly, V' is said to be *reachable* from V if there exists some path from V to V', denoted as $V \ll V'$. □

All vertices (hyperedges) reachable from a node V (hyperedge E) indicate the threat space induced from V (E). For example, E_d shown in Fig. 3(b) is reachable from E_a, E_b and E_c^2. The above described DBH has an important feature in that the number of vertices in a DBH is in general much smaller than the number of the corresponding TSS search nodes.

4   Methods of DB Search

In NCTU6, a threat-space search routine, named IsVCDT, is used to check whether the current position is a win for the attacker by VCDT within a limited depth. The routine is described as follows.

```
Procedure IsVCDT(Ψ, depth)
 1:    If the current position is winning return winning;
 2:    If depth=0 return not winning;
 3:    for each macro-move M do
 4:        if WillSearch (Ψ, M) then
 5:            let Ψ' be Ψ and append M into Ψ'
 6:            AddMove(M);
 7:            isWin = IsVCDT(Ψ', depth-1);
 8:            mark E(M) as inactive;
 9:            if isWin return winning;
10:        end if
11:    end for
12:    return not winning;
end
```

In lines 3 to 8 of IsVCDT, all candidate macro-moves are investigated in Will-Search (Ψ, M) to decide whether the macro-moves M are to be searched. Previously, in NCTU6, the routine WillSearch (Ψ, M) is in general true, except for some clearly bad moves or up to a limited number of moves. In this paper, the routine is modified to filter more macro-moves using dependency-based search techniques. Four methods for filtering are proposed in the following four subsections respectively.

4.1   Method F1

Method F1 only allows dependent moves to be searched. Given a CCDT $\Psi = \langle M_1, M_2, \ldots, M_{i-1} \rangle$ and the next candidate M_i, WillSearch (Ψ, M_i) returns true (i.e., M_i is to be searched) if and only if the following condition holds. For all M_j, $j \leq i$, there exists some hyperedge $E_j \in \mathbb{E}(M_j)$ such that $E_j \ll E_i$, for every $E_i \in \mathbb{E}(M_i)$.

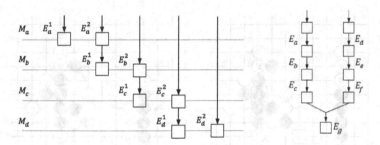

Fig. 4. (a) A DBH with 4 moves M_a, M_b, M_c, M_d, and (b) a DBH with a combination macro-move E_g.

In F1, we only traverse the moves which depend on their precedent moves and are reachable from all ancestors in order to reduce the number of nodes to search. Consider the DBH in Fig. 4(a), where a variety of sequences can be searched from the same eight primitive moves $E_a^1, E_a^2, E_b^1, E_b^2, E_c^1, E_c^2, E_d^1, E_d^2$ Examples include $\langle M_{aa}^{12}, M_{bb}^{12}, M_{cc}^{12}, M_{dd}^{12} \rangle$, $\langle M_{ab}^{12}, M_{cc}^{12}, M_{da}^{12}, M_{bd}^{12} \rangle$, $\langle M_{ac}^{12}, M_{da}^{12}, M_{bb}^{12}, M_{cd}^{12} \rangle$, etc., where M_{xy}^{mn} indicates a macro-move from E_x^m and E_y^n and M_x is M_{xx}^{12}. Using F1, the search will traverse from M_a to M_b, but since E_c^1 is not reachable from E_a^1, the search cannot continue past M_b.

Also, this method cannot be applied to cases where a combination move is required, such as the example in Fig. 4(b). Another example can be observed in Fig. 3(b), where WILLSEARCH($\langle M_b \rangle$, M_c) returns true, but WILLSEARCH($\langle M_b, M_c \rangle$, M_a) returns false, as does WILLSEARCH($\langle M_a \rangle$, M_b), since M_a and M_b are mutually independent. Thus, the winning sequences $\langle M_a, M_b, M_c, M_d \rangle$ and $\langle M_b, M_c, M_a, M_d \rangle$ cannot be found. Intuitively, F1 focuses purely on reducing the search size, but the reduction comes at the cost of being unable to find sequences requiring combinations.

4.2   Method F2

Method F2 allows, in addition to those in F1, macro-moves that can form a combination move later. Before discussing F2, we define two types of zones below.

Definition 4:   Each hyperedge E is associated with an *attacking zone*, denoted by $Z^A(E)$, which is a union set of locations of attacker pieces $P(E'^A)$ for all hyperedges E' reachable from E (i.e., $E \ll E'$). For each macro-move M, $Z * (M)$ denote the *to-be-combined zone*, which is a set of locations of the board which are on the lines containing one of the pieces in $P(M^A)$.   □

For example, in Fig. 5(a), the marked squares form the attacking zone $Z^A(E_b)$. Note that the locations of single-threats like G9, F10 and E11 are not in the zone since the threat does not depend on E_b. In Fig. 5(b), the to-be-combined zone $Z^*(M_a)$ is the collection of squares that are covered by red lines[2]. The intersection of both $Z^*(M_a) \cap Z^A(E_b)$ contains the locations E2 and F3. This implies that some of subsequent macro-moves from M_b may combine with M_a at E2 or F3.

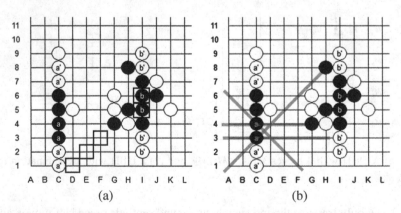

Fig. 5. An example of how combination detection is implemented. (a) $Z^A(E_b)$ (b) $Z^*(M_a)$ (Color figure online).

---

[2] In actual implementation for the zone $Z^*(M_a)$, all lines centered on $P(M_a^A)$ are not longer than 6 in all directions. Moreover, the lines are shortened when encountering opponent pieces. The zone is denoted by Z^*, since these lines form a star-like shape from the attacking pieces.

In Method F2, WILLSEARCH(Ψ, M_i) returns true if Method F1 returns true. Additionally, WILLSEARCH(Ψ, M_i) returns true if the following condition holds: there exists some $E_i \in \mathbb{E}(M_i)$ such that $Z^*(M_{i-1}) \cap Z^A(E_i) \neq \emptyset$. Thus, Method F2 allows TSS to switch to another move M_i whose descendants may be combined with the current move M_{i-1}. For example, in Fig. 5(a), since the intersection $Z^*(M_a) \cap Z^A(E_b)$ is not empty, the program is allowed to play M_b.

One issue is the sequence in which moves are played. Consider the example illustrated in Fig. 4(b). For the move E_g, which is a combination of E_c and E_f, Method F2 plays E_d after E_a, E_b, E_c, but does not play E_d immediately after E_a, E_b. This ensures that the playing sequence is narrowed-down to a unique sequence, eliminating the various different sequences that can also lead to E_g.

4.3 Method F3

Method F3 allows, in addition to the previous methods, moves that can block inversions. Inversions are defensive moves that result in the defender generating threat(s). To avoid inversions, the attacker needs to play moves that block defender threats while still generating a double-threat move of its own. Threat-blocking can be classified into post-blocking and pre-blocking. For post-blocking, the attacker blocks defender threats in the next immediate move, while for pre-blocking the attacker blocks the threats in advance.

In Method F3, WILLSEARCH(Ψ, M_i) returns true if F2 (including F1) returns true. Additionally, it returns true in the following ways. For post-blocking, i.e., M_{i-1} is an inversion, it is straightforward for WILLSEARCH(Ψ, M_i) to return true if M_i can block opponent threats while generating a double-threat for the attacker.

Before examining pre-blocking, we define the *defending-inversion zone*, denoted by $Z^I(E)$. Each hyperedge E is associated with a set of locations $Z^I(E)$ that potentially blocks defender threats $T(E'^D)$ for all hyperedges E' reachable from E (i.e., $E \ll E'$).

Assume that $P(M_{i-1}) \cap Z^I(E_i) \neq \emptyset$, where $E_i \in \mathbb{E}(M_i)$. Then, M_{i-1} may pre-block E_i's descendants that end up becoming defender threats. In Method F3, WILLSEARCH(Ψ, M_i) also returns true in this case for potential pre-blocking.

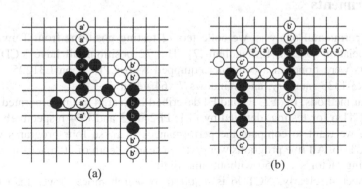

(a)   (b)

Fig. 6. (a) An example of post-block. (b) An example of pre-block.

The method is illustrated by two examples in Fig. 6, where the inversions are underlined. In Fig. 6(a), M_b will be searched since it blocks the inversion generated by M_a^D. In Fig. 6(b), M_b blocks the inversion generated by M_c^D in advance and we can then search M_c after M_b. If we searching M_c first, post-blocking is not possible since the attacker has to play two moves, M_a^A and M_b^A, to block the inversion. Therefore, the order in which moves are searched is critical for pre-blocking.

In actual implementation, we do not support a complete set for $Z^I(E)$. The larger $Z^I(E)$ is, the higher the chance to switch search order and the larger the branching factor. In our implementation, we only consider continuous inversions for $Z^I(E)$.

4.4   Method F4

Using the above three methods, we still encounter a problem: the number of moves derived from the three methods (even for Method F1) is still large in the case of macro-moves in T_{ST}^I. Method F4 is proposed to filter more redundant moves in T_{ST}^I.

Consider an example where many independent half-moves (single-threats), say $E_1, E_2, \ldots, E_n$, are available but only E_1 depends on the last move M_{i-1}. Based on F1, all macro-moves including E_1 and any other E_j are all legal candidate moves. In the case that all the other E_j are single-threats without any further development, the search still runs once for each of these E_j. However, it is clear for human players to search E_1 with some E_i only once.

In order to solve this problem, Method F4, basically following F3, is proposed to reduce the search space. Assume that F3 returns true and the candidate move M_i is in T_{ST}^I. For simplicity, let $\mathbb{E}(M_i) = \{E_1, E_2\}$, and E_1 depend on the last move M_{i-1} without loss of generality. Method F4 checks the following cases. First, in the case that the half-move for E_2 is actually not created in the DBH yet, WILLSEARCH(Ψ, M_i) returns true. Second, in the case that E_2 is in the DBH already, investigate the intersection of $Z^A(E_1)$ and $Z^A(E_2)$. If the intersection is not empty, then WILL-SEARCH(Ψ, M_i) returns true, since it is likely to combine both half-moves later. WILLSEARCH(Ψ, M_i) returns false otherwise.

5   Experiments

For our experimental analysis, we collected 72 testing positions from Taiwan Connect6 Association [9] and Littlegolem [7]. The 72 positions all have VCDTs. Our experiments were done on a computer equipped with CPU Xeon E31225 3.1 GHz, 4 GB DDR3-1333 memory, and Windows 7 x64 version.

The four methods, F1, F2, F3 and F4 described in Sect. 4 are incorporated into the original NCTU6 for testing, denoted by PF1, PF2, PF3 and PF4 respectively. In this subsection, we want to compare the performances of these four programs with the original NCTU6. All five programs were set to a search depth of 10 macro-moves with the branching factor set to 50 without time limits.

From Sect. 4, clearly, NCTU6 is required to search more moves than the four programs PF1 to PF4, PF3 is more than PF2, and PF2 is more than PF1. Hence, the

Table 1. The comparison among the five programs.

Programs	Move count	Time (S)	Solve positions
NCTU6	12,694,339	1,050	72
PF1	55,691	11	55
PF2	5,648,130	571	70
PF3	7,273,329	799	72
PF4	2,852,988	255	72

following are expected: NCTU6 is slower than PF1 to PF4, but more accurate; PF3 is slower than PF2, but more accurate; and PF2 is slower than PF1, but more accurate.

Table 1 shows the experimental results fit the above expectation. Program PF4 solved all 72 positions while performing better than NCTU6, PF2 and PF3. The average speedup of PF4 over the original NCTU6 was 4.12, and the speedup for some hard positions reached up to 50.3. Figure 7 (below) shows the computation times of all the 72 positions for NCTU6 and PF4. The position IDs are sorted in the order of the computation times for NCTU6.

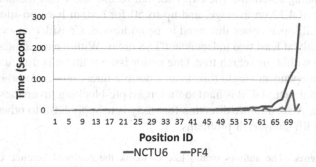

Fig. 7. The computation time on each position by original NCTU6 and NCTU6 with DBS.

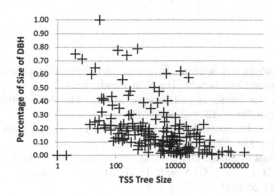

Fig. 8. The percentage of DBH over TSS tree size for all benchmark.

Table 1 also shows that the average computation time for each searched move is 82.7 ms in NCTU6, while it is 89.4 ms for PF4. This implies that the incurred overhead is about 8.1 % of the computation time.

Now, we want to investigate the memory size of DBH with respect to the TSS tree size. We use the number of hyperedges to indicate the memory size for DBH, and the number of searched moves to indicate the TSS tree size. Figure 8 shows the percentage of DBH size over TSS tree size for all testing positions. We can see that the ones with large TSS tree sizes usually have lower percentage of memory requirement for DBH.

6 Conclusion

For Connect6 and other k-in-a-row games, DBS is critical for TSS when there are many independent or nearly independent TSS areas, a situation that is especially common during end game. Unfortunately, the rule that two pieces are played per move in Connect6 makes DBS extremely difficult to apply to Connect6 programs.

This paper is the first attempt to apply DBS to Connect6 programs. We propose four DBS methods, F1 to F4, to reduce the search space of TSS in NCTU6 while still successfully finding solutions. The experimental results show that method F4 yields a speedup factor of 4.12 on average, and up to 50 for certain hard positions.

There are still open issues that need to be addressed. Consider an example where there are currently at least two independent TSS areas. Within each area, single threats may exist deep within the search tree. One major issue is that it is difficult to associate two of these independent single-threats as a double-threat move. Another issue is, as pointed out in Subsect. 4.4, it is hard to find more pre-blocking inversions without loss of performance. Also, a future direction for study is to apply DBS to other games such as Go to solve life-and-death problems.

Acknowledgements. The authors would like to thank the National Science Council of the Republic of China (Taiwan) for financial support of this research under contract numbers NSC 97-2221-E-009-126-MY3, NSC 99-2221-E-009-102-MY3 and NSC 99-2221-E-009-104-MY3.

References

1. Allis, L.V.: Searching for solutions in games and artificial intelligence. Ph.D. Thesis, University of Limburg, Maastricht, The Netherlands (1994)
2. Allis, L.V., van den Herik, H.J., Huntjens, M.P.H.: Go-Moku solved by new search techniques. Comput. Intell. **12**, 7–23 (1996)
3. van den Herik, H.J., Uiterwijk, J.W.H.M., Rijswijck, J.V.: Games solved: now and in the future. Artif. Intell. **134**, 277–311 (2002)
4. ICGA (International Computer Games Association). http://ticc.uvt.nl/icga/
5. Lin, H.-H., Sun, D.-J., Wu, I.-C., Yen, S.-J.: The 2010 TAAI computer-game tournaments. ICGA J. **34**(1), 51–55 (2011)

6. Lin, P.-H., Wu, I.-C.: NCTU6 wins in the Man-Machine Connect6 championship 2009. ICGA J. (SCI) **32**(4), 230–233 (2009)
7. Little Golem website. http://www.littlegolem.net/
8. Renju International Federation: The International Rules of Renju. http://www.renju.net/study/rifrules.php
9. Taiwan Connect6 Association: Connect6 Homepage. http://www.connect6.org/
10. TCGA Association: TCGA Computer Game Tournaments. http://tcga.ndhu.edu.tw/TCGA2011/
11. Allis, Private Communication (2012)
12. Thomsen, T.: Lambda-search in game trees - with application to Go. ICGA J. **23**(4), 203–217 (2000)
13. Wu, I.-C., Huang, D.-Y., Chang, H.-C.: Connect6. ICGA J. **28**(4), 234–242 (2006)
14. Wu, I.-C., Huang, D.-Y.: A new family of k-in-a-row games. In: The 11th Advances in Computer Games Conference (ACG'11), pp. 180–194, Taipei, Taiwan, (2005)
15. Wu, I.-C., Lin, P.-H.: NCTU6-Lite wins Connect6 tournament. ICGA J. **31**(4), 240–243 (2008)
16. Wu, I.-C., Lin, P.-H.: Relevance-Zone-Oriented proof search for Connect6. IEEE Trans. Comput. Intell. AI Games (SCI) **2**(3), pp. 191–207 (2010)
17. Wu, I.-C., Yen, S.-J.: NCTU6 wins Connect6 tournament. ICGA J. **29**(3), 157–158 (2006)
18. Yen, S.-J., Yang, J.-K.: 2-Stage Monte Carlo tree search for Connect6. IEEE Trans. Comput. Intell. AI Games (SCI) **3**(2), pp. 100–118, ISSN: 1943–068X, (2011). doi:10.1109/TCIAIG.2011.2134097

On Semeai Detection in Monte-Carlo Go

Tobias Graf, Lars Schaefers[(✉)], and Marco Platzner

University of Paderborn, Paderborn, Germany
tobiasg@mail.uni-paderborn.de, {slars,platzner}@uni-paderborn.de

Abstract. A frequently mentioned limitation of Monte-Carlo Tree Search (MCTS) based Go programs is their inability to recognize and adequately handle capturing races, also known as *semeai*, especially when many of them appear simultaneously. The inability essentially stems from the fact that certain group status evaluations require deep lines of correct tactical play which is directly related to the exploratory nature of MCTS. In this paper we provide a technique for heuristically detecting and analyzing semeai during the search process of a state-of-the-art MCTS implementation. We evaluate the strength of our approach on game positions that are known to be difficult to handle even by the strongest Go programs to date. Our results show a clear identification of semeai and thereby advocate our approach as a promising heuristic for the design of future MCTS simulation policies.

1 Introduction

Monte-Carlo Tree Search (MCTS) is a class of simulation-based search algorithms that brought about great success in the past few years regarding the evaluation of stochastic and deterministic two-player games such as the Asian board game Go. MCTS is a simulation-based algorithm that learns a value function for game states by consecutive simulation of complete games of self-play using randomized policies to select moves for either player. MCTS may be classified as a sequential best-first search algorithm [12] that uses statistics about former simulation results to guide future simulations along the search space's most promising paths in a best-first manner. A crucial part of this class of search algorithms is the playout policy used for move decisions in game states where statistics are not yet available. The playout policy drives most of the move decisions during each simulation. Moreover, the simulations' final positions are the main source of data for all remaining computations. Accordingly, for the game of Go there exists a large number of publications about the design of such policies, e.g., [3,4,7]. One of the objectives that playout designers pursue focuses on simulation balancing to prevent biased evaluations [7,8,13]. Simulation balancing targets at ensuring that the policy generates moves of equal quality for both players in any situation. Hence, adding domain knowledge to the playout policy for attacking also necessitates adding domain knowledge for defending. One of the greatest early improvements in Monte-Carlo Go was the idea of sequence-like playout policies [7] that highly concentrate on local answer moves. They lead to a

H.J. van den Herik et al. (Eds.): CG 2013, LNCS 8427, pp. 14–25, 2014.
DOI: 10.1007/978-3-319-09165-5_2, © Springer International Publishing Switzerland 2014

highly selective search. Further concentration on local attack and corresponding defense moves ameliorated the handling of some types of semeai and hereby contributed to additional strength improvement of MCTS programs. However, by adding more and more specific domain knowledge with the result of increasingly selective playouts, the door is opened for more imbalance. This in turn allows for severe false estimates of position values. Accordingly, the correct evaluation of semeai is still considered to be quite challenging for MCTS-based Go programs [11]. This holds true, especially when they require long sequences of correct play by either player. In order to face this issue, we search for a way to make MCTS searchers become aware of probably biased evaluations due to the existence of semeai or groups with uncertain status. In this paper we present our results about the analysis of score histograms to infer information about the presence of semeai. We heuristically locate the fights on the Go board and estimate their corresponding relevance for winning the game. The developed heuristic is not yet used by the MCTS search. Accordingly, we cannot definitely specify and empirically prove the benefit of the proposed heuristic in terms of playing strength. All experiments are performed with our MCTS Computer Go engine GOMORRA. The main contributions of this paper are as follows.

- Analysis of score histograms towards the presence of semeai during an MCTS search process.
- Stochastic mapping of score clusters derived from histograms to board sites of individual semeai.
- Experimental evaluation of our approach on a variety of test positions that are known to be difficult to handle by modern MCTS-based Go programs.

In Sect. 2 we give a summary of related work and some background about clustering and mode finding in empirical density functions and about the meaning and computation of the Monte-Carlo (MC) criticality measure. Section 3 details our concrete method for the detection of semeai by the analysis of score histograms. The method was used for our experiments presented in Sect. 4. Finally, in Sect. 5 we draw conclusions and give directions for future work.

2   Background and Related Work

A central role in our proposed method for identifying the presence and concrete location of semeai in game positions, takes into account the clustering of MC simulations into groups related to the score they achieve in their corresponding terminal positions. We decided to use a mean shift algorithm on score histograms for this task. Mean shift is a straightforward mode-seeking algorithm that was initially presented in [6] and more deeply analyzed in [2]. It is derived from a gradient ascent procedure on an estimated density function that can be generated by Kernel Density Estimation on sampled data. Modes are computed by iteratively shifting the sample points to the weighted mean of their local neighboring points until convergence. Weighting and locality are hereby controlled by a kernel and a bandwidth parameter. We obtain a clustering by assigning each point in the sample space the nearest mode computed by the mean shift algorithm. See Subsect. 3.2 for formulae and detailed information about our implementation.

The stochastic mapping of clusters to relevant regions on the Go board is realized by the application of a slightly modified MC-criticality measure. Starting in 2009 several researchers came up with the idea of using an intuitive covariance measure between controlling a certain point of a Go board and winning the game. The covariance measure is derived from the observation of MC simulation's terminal positions [5][1] [1,10]. Such kind of measures are most often called *(MC-)Criticality* when used in the context of Computer Go. The particular publications around this topic differ slightly in the exact formula used but foremost in the integration with the general MCTS framework. Coulom [5] proposed the use of MC-Criticality as part of the feature space of a move prediction system that in turn is used for playout guidance in sparsely sampled game states, while Pellegrine et al. [10] used the criticality measure with an additional term in the UCT formula. Baudis and Gailly [1] argued for a correlation of MC-Criticality and RAVE and consequently integrated both.

3   MC-Criticality Based Semeai Detection

In this section, we present our approach for detecting and localizing capturing races (jap.: semeai) in positions of the game of Go by the clustering of MC simulations according to their respective score and the computation of cluster-wise MC-criticality. When performing an MCTS search on a game position, a number of randomized game continuations are generated from the position under consideration. Each of these simulations ends in a terminal game position that can be scored according to the game rules. In case of Go, the achievable score[2] per simulation ranges from -361 to $+361$. A common first step for obtaining information about the score distribution is the construction of a score histogram that can be interpreted as an empirical density function by appropriate scaling. Assuming that the presence of semeai likely results in more than one cluster of simulation scores (depending on whether the one or the other player wins) we are interested in identifying such clusters and the corresponding regions on the Go board that are responsible for each particular cluster. Accordingly, semeai detection with our approach is limited to cases in which different semeai outcomes lead to distinguishable final game scores. In the following, we first introduce some notations and afterwards step through our method starting with clustering.

3.1   Notations

Let $S \subseteq \mathbb{Z}$ be the discrete set of achievable scores. Having built a histogram H of a total of n simulation outcomes, we write $H(s)$ for the number of simulations

---

[1] Available online: http://remi.coulom.free.fr/Criticality/

[2] The player that makes the second move in the game is typically awarded some fixed bonus points, called komi, to compensate for the advantage the other player has by making the first move. Typical komi values are 6.5, 7 and 7.5, depending on the board size. Accordingly, the score range might become asymmetric.

that achieved a final score of $s \in S$, hence $\sum_{s \in S} H(s) = n$. We denote the average score of n simulations by $\bar{s} = \sum_{s \in S} H(s)/n$. Each element c of the set of score clusters C is itself an interval of scores, hence $c \subseteq S$. All clusters are disjunct in respect to the score sets they represent. We write $c(s)$ for the single cluster to which a score s is assigned to.

3.2   Clustering

As mentioned in Sect. 2 we use a mean-shift algorithm for mode seeking and clustering in score histograms that depends on the choice of a kernel and a parameter h that is called bandwidth. For our implementation we use the triweight kernel K defined as

$$K(x) = \frac{35}{32} \left(1 - x^2\right)^3 \mathbb{I}(|x| \leq 1),$$

where $\mathbb{I}$ denotes the indicator function that equals to 1 if the condition in the argument is true and to 0 otherwise. Figure 1 shows a plot of the kernel. We use *Silverman's rule of thumb* [14] (p. 48) to determine the bandwidth parameter h based on the number of simulations n and their variance:

$$h = 3.15 \cdot \hat{\sigma} n^{-\frac{1}{5}} \quad \text{with } \hat{\sigma}^2 = \frac{\sum_{s \in S}(s - \bar{s})^2 H(s)}{n - 1}.$$

When perceiving $f(s) = H(s)/n$ as an empiric density function, we can use Kernel Density Estimation with the triweight kernel and the computed bandwidth parameter to obtain a smooth estimated density function $\hat{f}$:

$$\hat{f}(y) = \frac{1}{nh} \sum_{s \in S} H(s)K\left(\frac{y - s}{h}\right).$$

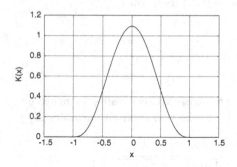

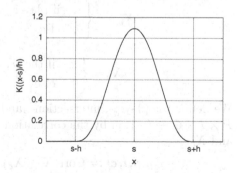

Fig. 1. The Triweight kernel $K(x)$ **Fig. 2.** $K\left(\frac{x-s}{h}\right)$ as used in $\hat{f}(y)$

A plot of the term $K\left(\frac{x-s}{h}\right)$ is shown in Fig. 2. An example for a normalized histogram and the corresponding estimated density function $\hat{f}$ can be found, e.g., in Fig. 3b, where $\hat{f}$ is represented by the black curve.

To find the modes of $\hat{f}$, we initialize a mode variable m_s for each score $s \in S$ to the respective score itself, i.e., $m_s = s$. The mean shift algorithm now iteratively updates this mode variables by

$$m_s = \frac{\sum_{s \in S} H(s) K\left(\frac{m_s - s}{h}\right) s}{\sum_{s \in S} H(s) K\left(\frac{m_s - s}{h}\right)}$$

until convergence. The different conversion points of the mode variables are the positions of the modes of $\hat{f}$. We build one score cluster for each mode and write m_c for the position of the mode that corresponds to cluster c. To account for estimation errors and sample variances we only consider clusters corresponding to mode positions m with $\hat{f}(m) \geq T$ for some appropriate threshold T. Each score s is then assigned to the cluster $c(s) = \operatorname{argmin}_{c \in C} |m_s - m_c|$.

3.3   Player-, Intersection-, and Cluster-Wise MC-Criticality

MC-criticality was introduced as an intuitive covariance measure between controlling a certain intersection of a Go board and winning the game. Here, controlling an intersection from the viewpoint of a specific player means that in the game's terminal position the intersection is either occupied by a stone of the player, or the intersection is otherwise counted as the players territory (e.g. builds an eye of the player). We propose a slightly modified measure to compute the correlation of controlling a point on the Go board and achieving a score that falls into a given interval. Let $P = \{black, white\}$ be the set of players, I the set of all board intersections and C a set of score clusters determined as presented in Subsect. 3.2. Given a number of terminal game positions generated by MCTS simulations, we define the following random variables.

$$X_{p,i} = \begin{cases} 1, & \text{if player } p \text{ controls intersection } i \\ 0, & \text{else} \end{cases}$$

$$X_c = \begin{cases} 1, & \text{if the score falls into cluster } c \\ 0, & \text{else} \end{cases}$$

We define the player-, intersection-, and cluster-wise MC-criticality measure $g :$ $P \times I \times C \to (-1, 1)$ by the correlation between the two random variables $X_{p,i}$ and X_c:

$$g(p, i, c) := \operatorname{Corr}(X_{p,i}, X_c) = \frac{\operatorname{Cov}(X_{p,i}, X_c)}{\sqrt{\operatorname{Var}(X_{p,i})}\sqrt{\operatorname{Var}(X_c)}}$$

which gives

$$g(p, i, c) = \frac{\mu_{p,i,c} - \mu_{p,i}\mu_c}{Z} \quad \text{with } Z = \sqrt{\mu_{p,i} - \mu_{p,i}^2}\sqrt{\mu_c - \mu_c^2},$$

for $\mu_{p,i,c} = E[X_{p,i}X_c]$, $\mu_{p,i} = E[X_{p,i}]$ and $\mu_c = E[X_c]$ with $E[\cdot]$ denoting the expectation. Accordingly, $\boldsymbol{\mu_{p,i,c}}$ denotes the ratio of all n simulations' terminal positions in which player p controls intersection i and the score falls into cluster c, $\boldsymbol{\mu_{p,i}}$ represents the ratio of the simulations' terminal positions in which player p controls intersection i regardless of the score and $\boldsymbol{\mu_c}$ is the ratio of simulations with a final score that falls into cluster c. The measure $g(p,i,c)$ gives the criticality for player p to control intersection i by the end of the game in order to achieve some final score $s \in c$. The lowest possible value that indicates a highly negative correlation is -1. Here, negative correlation means that it is highly unlikely to end up in the desired score cluster if player p finally controls intersection i. Our measure becomes most similar to the former published intersection-wise criticality measures when choosing the cluster $C_{\mathrm{black}} = \{s \in S | s > 0\}$ representing a black win, and $C_{\mathrm{white}} = \{s \in S | s < 0\}$ representing a white win. This clustering then resembles the criticality by

$$g_{\mathrm{former}}(i) \approx g(\mathrm{black}, i, C_{\mathrm{black}}) + g(\mathrm{white}, i, C_{\mathrm{white}})$$

with the difference that $g(p,i,c)$ uses the correlation instead of the covariance.

3.4   Detecting and Localizing Semeai

Putting the clustering procedure from Subsect. 3.2 and the criticality measure of Subsect. 3.3 together, we obtain a method (1) for analyzing complete board positions with respect to a possible presence of semeai and (2) in case they exist, the method even allows for approximately localizing them on the board. To analyze a given position, we perform standard MCTS and collect data about the simulations' terminal positions which is necessary to derive later on the score histogram H and the values for $\mu_{p,i,c}$, μ_c and $\mu_{p,i}$. All we need for this purpose is (1) a three-dimensional array *control* with a number of $|P| \cdot |I| \cdot |S|$ elements of a sufficiently large integer data type, (2) initialize all elements to zero and (3) increment them appropriately at the end of each MC simulation. Here, for each terminal position the value of element $control(p,i,s)$ is incremented in case player p controls intersection i and the position's score equals s. Given this relation, we derive the histogram function H by

$$H(s) = \sum_{p \in P} control(p,i,s) \quad \text{for some fixed } i \in I.$$

Having the score histogram of $n = \sum_{s \in S} H(s)$ simulations, we apply the clustering procedure as described in Subsect. 3.2 to obtain the set of score clusters C. As mentioned in Subsects. 3.1 and 3.2, each cluster $c \in C$ is constructed around a mode of $\hat{f}$ and we denote the corresponding mode's position by m_c. Given this, we can derive the values for $\mu_{p,i,c}$, μ_c, and $\mu_{p,i}$:

$$\mu_{p,i,c} = \frac{1}{n} \sum_{s \in c} control(p,i,s),$$

$$\mu_c = \frac{1}{n} \sum_{s \in c} H(s),$$

$$\mu_{p,i} = \frac{1}{n} \sum_{s \in S} \text{control}(p, i, s).$$

This in turn allows for the cluster-wise criticality computation as described in Subsect. 3.3, that, for each player determines the criticality of controlling the corresponding intersection in order to make the game end with a score belonging to the respective cluster. In case more than one cluster was found, the resulting distribution of criticality values for a given player and cluster, typically shows high valued regions that are object to a semeai. Thereby, the criticality values represent a stochastic mapping of each cluster to board regions with critical intersections that have to be controlled by one player in order to achieve a score that corresponds to the cluster. By further comparison of the critical board regions of the varying clusters and under consideration of the clusters mode positions m_c, it might even be possible to estimate the value of a single semeai in terms of scoring points. In the next section, we present results achieved with our approach on a number of example positions.

4   Experimental Results

Based on our Go program GOMORRA, we implemented our approach and made a series of experiments on different Go positions that contain multiple semeai. For our experiments, we concentrated on a number of two-safe-groups test cases out of a regression test suite created by Shih-Chieh (Aja) Huang (6d) and Martin Müller [9]. The collection of problems in this test suite was especially created to reveal the weaknesses of current MCTS Go engines and is part of a larger regression test suite of the FUEGO Open Source project[3].

Figure 3 shows one of the test positions that contains two semeai, one on the upper right, the other on the lower right of the board. Black is to play and the result should be a clear win for white, hence a negative final score, because both semeai can be won by the white player. Figure 3b shows the corresponding score histogram of 128,000 MC simulations. The colors indicate the clustering computed by the method described in Subsect. 3.2. Figure 3a, c, d, and e show the respective criticality values for the white player to end up in cluster 1, 2, 3, and 4, counted from left to right. Positive correlations are illustrated by white squares of a size corresponding to the degree of correlation. Each intersection is additionally labeled with the criticality value. One can clearly see how the different clusters map to the different possible outcomes of the two semeai.

Figure 4a–d show the results for test cases 1, 2, 3, and 5 of the above mentioned test suite (test case 4 is already shown in Fig. 3). Due to space limitations, we restrict the result presentation to the score histograms and the criticality for player white to achieve a score assigned to the leftmost cluster.

---

[3] See: http://fuego.svn.sourceforge.net/viewvc/fuego/trunk/regression/

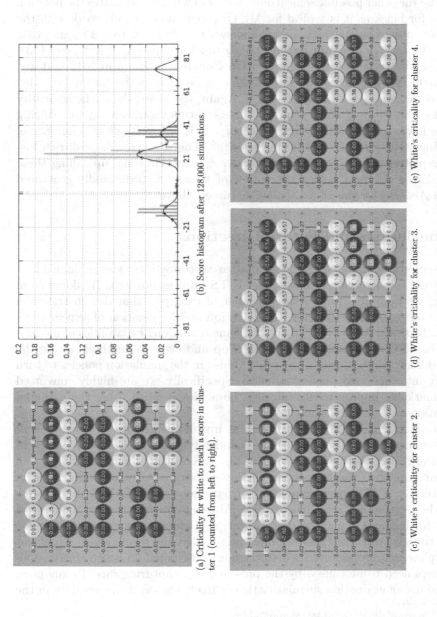

(a) Criticality for white to reach a score in cluster 1 (counted from left to right).

(b) Score histogram after 128,000 simulations.

(c) White's criticality for cluster 2.

(d) White's criticality for cluster 3.

(e) White's criticality for cluster 4.

Fig. 3. The player-wise criticality values reveal the critical intersections on the board for making the game end with a score associated to a specific cluster. The locations of the two capturing races are clearly identified (Color figure online).

The histograms and criticality values always show the correct identification of two separate semeai. We took the leftmost cluster, as for the given test cases this is the one containing the correct evaluation, i.e., a win for the white player. As can be derived from the shown histograms, in all cases, GOMORRA gets distracted by the other possible semeai outcomes and wrongly estimates the position as a win for black as it is typical for MCTS programs that only work with the mean outcome of simulations. Figure 5 shows results for a 13 × 13 game that was lost by the Go program PACHI playing black against Alex Ketelaars (1d). The game was played at the European Go Congress in Bonn in 2012 and was one of only two games lost by PACHI. As can be seen in the histogram, Alex managed to open up a number of semeai. Again, the board shows the criticality values for the leftmost cluster and reveals GOMORRA's difficulties in realizing that the lower left is clearly white's territory. It is one example of a number of games Ingo Althöfer collected on his website[4]. The site presents peak-rich histograms plotted by the Go program CRAZY STONE. He came up calling them *Crazy Shadows*. Rémi Coulom, the author of CRAZY STONE, kindly generated a *Crazy Analysis* for the 13 × 13 game discussed above[5].

5   Conclusions and Future Directions

We presented a method to detect and localize capturing races and explained how to integrate the detection into existing MCTS implementations. By doing so in practice, we were able to present a number of examples that demonstrate the power of our approach. However, the detection and localization of semeai alone is only a first step towards improving the semeai handling capabilities of modern MCTS-based Go programs. We must develop and evaluate methods to use the gathered knowledge in form of criticality values in the simulation policies to turn it finally into increased playing strength. Specifically, we are highly convinced that remarkable improvements can only be achieved when using the gathered information even in the playout policies.

Accordingly, in future work, we plan to investigate the use of some kind of large asymmetric shape patterns that dynamically adapt their size and shape to the critical regions as they are determined by the presented method. Integrating those patterns into existing move prediction systems as they are widely used in Computer Go in addition to training their parameters during the search process builds the next interesting challenge [15]. Already now, the results might be of interest for human Go players using Go programs to analyze game positions[6].

For the sake of correctness, we must admit that the term *semeai* might not be completely appropriately used throughout this paper. A score cluster does not always need to be caused by the presence of a capturing race. In any case, it represents an evaluation singularity that is likely caused by uncertainty in the

---

[4] http://www.althofer.de/crazy-shadows.html
[5] http://www.grappa.univ-lille3.fr/~coulom/CrazyStone/pachi13/index.html
[6] Some more context to existing visualizations for computer aided position analysis can be found online at http://www.althoefer.de/k-best-visualisations.html.

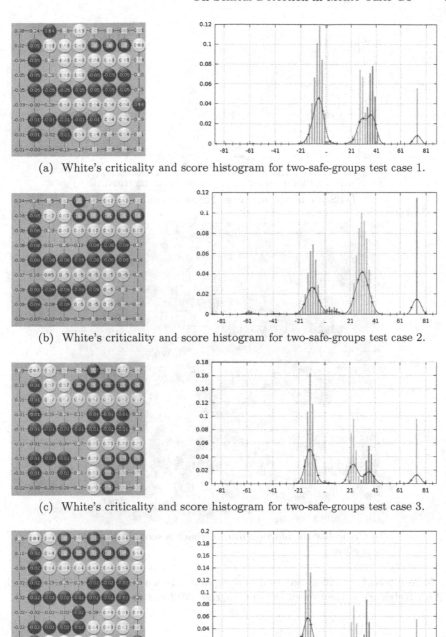

(a) White's criticality and score histogram for two-safe-groups test case 1.

(b) White's criticality and score histogram for two-safe-groups test case 2.

(c) White's criticality and score histogram for two-safe-groups test case 3.

(d) White's criticality and score histogram for two-safe-groups test case 5.

Fig. 4. The player-wise criticality values for another 4 test cases of the two-safe-groups regression test suite. The boards always show the criticality values for player white to end up with a score associated with the leftmost cluster of the corresponding histogram shown to the right. The numbering of test cases is as given in the test suite.

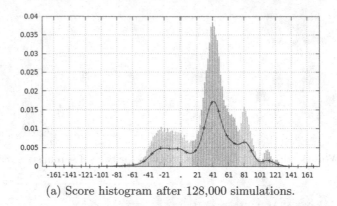

(a) Score histogram after 128,000 simulations.

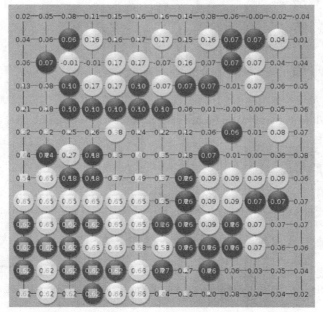

(b) Criticality for player white to reach a score of the left-most cluster.

Fig. 5. Analysis of a position occurred in the game between Go program PACHI (black) vs. Alex Ketelaars at the European Go Congress 2012 in Bonn (move 90).

evaluation of the life and death state of one or more groups of pieces. Also in this case our approach will help to localize the respective groups on the board.

Acknowledgements. We like to thank Ingo Althöfer for pointing the community to the potential of score histograms and thereby encouraging our work, as well as for kindly commenting on a preliminary version of this paper. We also thank Rémi Coulom for comments on an early version of this paper as well as for kindly providing an automated analysis by his Go program CRAZY STONE of one of the discussed example positions. We further thank Shih-Chieh (Aja) Huang and Martin Müller for the creation and sharing of their regression test suite.

References

1. Baudiš, P., Gailly, J.: PACHI: state of the art open source Go program. In: van den Herik, H.J., Plaat, A. (eds.) ACG 2011. LNCS, vol. 7168, pp. 24–38. Springer, Heidelberg (2012)
2. Cheng, Y.: Mean shift, mode seeking, and clustering. IEEE Trans. Pattern Anal. Mach. Intell. **17**(8), 790–799 (1995)
3. Coulom, R.: Efficient selectivity and backup operators in Monte-Carlo tree search. In: van den Herik, H.J., Ciancarini, P., Donkers, H.H.L.M.J. (eds.) CG 2006. LNCS, vol. 4630, pp. 72–83. Springer, Heidelberg (2007)
4. Coulom, R.: Computing Elo ratings of move patterns in the game of Go. ICGA J. **30**, 198–208 (2007)
5. Coulom, R.: Criticality: a Monte-Carlo heuristic for Go programs. Invited Talk at the University of Electro-Communications, Tokyo, Japan (2009)
6. Fukunaga, K., Hostetler, L.D.: The estimation of the gradient of a density function, with applications in pattern recognition. IEEE Trans. Inf. Theory **21**(1), 32–40 (1975)
7. Gelly, S., Wang, Y., Munos, R., Teytaud, O.: Modifications of UCT with patterns in Monte-Carlo Go. Technical report 6062, INRIA (2006)
8. Huang, S.-C., Coulom, R., Lin, S.-S.: Monte-Carlo simulation balancing in practice. In: van den Herik, H.J., Iida, H., Plaat, A. (eds.) CG 2010. LNCS, vol. 6515, pp. 81–92. Springer, Heidelberg (2011)
9. Huang, S.-C., Müller, M.: Investigating the limits of Monte Carlo tree search methods in computer Go. In: van den Herik, H.J., Iida, H., Plaat, A. (eds.) CG 2013. LNCS, vol. 8427. Springer, Heidelberg (2014)
10. Pellegrino, S., Hubbard, A., Galbraith, J., Drake, P.D., Chen, Y.-P.: Localizing search in Monte-Carlo go using statistical covariance. ICGA J. **32**(3), 154–160 (2009)
11. Rimmel, A., Teytaud, O., Lee, C.-S., Yen, S.-J., Wang, M.-H., Tsai, S.-R.: Current frontiers in computer Go. IEEE Trans. Comput. Intell. AI Games **2**, 229–238 (2010)
12. Silver, S.: Reinforcement learning and simulation-based search in computer Go. Ph.D. thesis, University of Alberta (2009)
13. Silver, D., Tesauro, G.: Monte-Carlo simulation balancing. In: International Conference on Machine Learning, pp. 945–952 (2009).
14. Silverman, B.W.: Density Estimation For Statistics and Data Analysis. Monographs on Statistics and Applied Probability, vol. 26. Chapman & Hall/CRC, London (1986)
15. Wistuba, M., Schaefers, L., Platzner, M.: Comparison of Bayesian move prediction systems for computer Go. In: Proceedings of the IEEE Conference on Computational Intelligence and Games (CIG), pp. 91–99, September 2012

Efficiency of Static Knowledge Bias in Monte-Carlo Tree Search

Kokolo Ikeda and Simon Viennot[✉]

Japan Advanced Institute of Science and Technology, Nomi, Japan
{kokolo,sviennot}@jaist.ac.jp

Abstract. Monte-Carlo methods are currently the best known algorithms for the game of Go. It is already shown that Monte-Carlo simulations based on a probability model containing static knowledge of the game are more efficient than random simulations. Some programs also use such probability models in the tree search policy to limit the search to a subset of the legal moves or to bias the search. However, this aspect is not so well documented. In this paper, we describe more precisely how static knowledge can be used to improve the tree search policy. We show experimentally the efficiency of the proposed method by a large number of games played against open source Go programs.

1 Introduction

Monte-Carlo methods have been applied to the game of Go since 1993 when Brügmann [1] introduced the method. Since the discovery of the Upper Confidence Tree (UCT) in 2006 by Kocsis and Szepesvari [2], Monte-Carlo methods are now the mainstream algorithm of all computer Go programs.

Initial Monte-Carlo methods used *random* simulations, but it soon became obvious that better results were obtained when static knowledge is used as a policy to obtain *realistic* simulations. In particular, in 2007, Coulom introduced the Bradley-Terry model to learn automatically the weights of features associated to the moves of the game [3]. Several of the strongest Go programs are now using this kind of model, with various detailed features.

Static knowledge has also been applied to the UCT part of the Monte-Carlo tree search. Here, Coulom [3] used the Bradley-Terry model to limit the number of searched moves, a method called *progressive widening*. Chaslot et al. [8], and also Huang [10] introduced a bias in the UCT algorithm to take into account the static knowledge.

Yet, despite the widespread use of these ideas of (1) progressive widening and (2) knowledge bias, the literature on the subject is only partial and does not show detailed experimental data. Our goal in this paper is to give a more detailed description of these two ideas, and show experimental evidence that they can be implemented and tuned efficiently by a Bradley-Terry model.

We also investigate as precisely as possible the effect of the parameters that we use in progressive widening and knowledge bias. In order to obtain a significant amount of data, we have performed most of our experiments on Go boards

H.J. van den Herik et al. (Eds.): CG 2013, LNCS 8427, pp. 26–38, 2014.
DOI: 10.1007/978-3-319-09165-5_3, © Springer International Publishing Switzerland 2014

of size 13×13. A total of 31,200 games were played against FUEGO and PACHI, two open-source programs.

The paper is organized as follows. First, in Sect. 2, we give a summary of Monte-Carlo Tree Search. Then, in Sect. 3, we describe the Bradley-Terry model and how it can be applied to evaluate statically the moves of the game and to improve the Monte-Carlo simulations. This is mainly a summary of the work by Coulom [3]. Then, in Sect. 4, we describe how this Bradley-Terry model can also be used to limit the number of searched moves, and we show significant experimental data of the effect. In Sect. 5, we describe how the same Bradley-Terry model can be used in a third way, as a bonus bias in the UCB formula. Finally, in Sect. 6, we discuss some other characteristics of our program that are possibly related to the efficiency of the proposed bias.

2 Monte-Carlo Tree Search

The main idea of Monte-Carlo Tree Search (MCTS) is to construct a search tree step by step by repeatedly choosing the most promising child node of the already searched tree, expanding it, and then evaluating the new leaf with random self-play simulations, often called *playouts*. The result of the simulations, usually a binary win or loss value, is back-propagated in all the parent nodes up to the root, and a new most promising leaf node is chosen.

2.1 UCT Algorithm

The common policy used in the search tree to select the most promising nodes is the UCT algorithm [2]. From a given parent node of the tree, it consists in choosing the child j with the largest μ_j value, defined as follows:

$$\mu_j = \frac{w_j}{n_j} + C \cdot \sqrt{\frac{\ln n}{n_j}} \qquad (1)$$

where n is the number of times the parent node has been visited, n_j the number of times the child j has been visited, w_j the number of times the child j leads to a winning simulation, and C a free parameter to balance the two terms.

The first term, called *exploitation*, is the expected winning ratio when the child j is selected, so that UCT tends to search more the children with a high winning ratio. The second term, called *exploration*, is a measure of the "uncertainty" of the winning ratio, so that UCT tends to search more the children that have not been visited much.

The success of the UCT algorithm is directly related to the balance between the two terms: when a child has a high winning ratio, it will be searched more than other children, but then, its exploration term will decrease, up to a point where another child with a bigger exploration term will be chosen. There is no theoretical ideal value for the parameter C, which needs to be tuned experimentally. A typical value in our program is $C = 0.9$.

Several improvements of the plain UCT algorithm have been developed. An important one is the Rapid Action Value Estimation (RAVE), described for example in [9]. This is an application of the all-moves-as-first (AMAF) heuristic already introduced in the first Monte-Carlo Go program in 1993 [1]. The idea is to propagate a simulation result not only to parent nodes but also to sibling nodes corresponding to the moves played in the simulation. It is useful to boost the startup phase of the search with noisy but immediately available values.

We have implemented the UCT algorithm in a Go program called NOMITAN. It relies mainly on the Bradley-Terry model described in the next section. Also, as discussed later in Subsects. 6.1 and 6.3, the RAVE heuristic is not used by NOMITAN. In 2013, NOMITAN reached a rank of 2 dan on the KGS server, under the account NoMiBot.

2.2   Experimental Settings

Two strong open-source Go programs, FUEGO [7] and PACHI [11,12] have been published. We have used FUEGO version 1.1 and PACHI version 9.01 (both published in 2011) as strong opponents in our experiments. These versions of FUEGO and PACHI both use the UCT algorithm with the RAVE heuristic, and without the Bradley-Terry model described in the next section.

In order to limit the computing time, the main experiments of this paper were performed on boards of size 13×13 with 2 s of thinking time per move on 12-core machines. On 13×13 boards, our approach leads to games of good quality, even with this short thinking time. At the start of the game, FUEGO reaches around 140,000 playouts, PACHI around 170,000 and our program around 19,000. Thus our playouts are fairly slow compared to FUEGO and PACHI, mainly because we use the Bradley-Terry model of the next section with relatively large patterns.

For all the experiments on the 13×13 boards, we have performed at least 400 games for each data point, which gives a 95 % confidence interval of less than ± 5 % on the winning rate results of the experiments. For the sake of readability, error bars are plotted only in Fig. 3.

3   Bradley-Terry Model

In 2007, Coulom applied the Bradley-Terry model to the static evaluation of the moves of the game of Go [3]. It is now widely used by other strong Go programs, such as ZEN [6], ERICA [10] or AYA [13].

3.1   Description of the Bradley-Terry Model

The Bradley-Terry model assigns a γ value, reflecting the strength, to each individual of a competition. The result of a competition between two individuals i and j can be predicted by the following probability.

$$P(i\ wins\ against\ j) = \frac{\gamma_i}{\gamma_i + \gamma_j}$$

The Bradley-Terry model extends well to competitions between several individuals. For example, in a competition between n individuals, the probability that i wins is given by $P(i\ wins) = \frac{\gamma_i}{\gamma_1+\gamma_2+...+\gamma_n}$.

In the most general case, we can consider a competition between teams of individuals. In this case, the strength of a team is given by the product of the γ values of the individuals in the team. For example, the probability that the team 1-2 wins in a competition involving the teams 1-2 (team composed of individuals 1 and 2), 3-4-5 and 1-6 is given by:

$$P(\text{1-2 } wins \ against \text{ 3-4-5 } and \text{ 1-6}) = \frac{\gamma_1\gamma_2}{\gamma_1\gamma_2+\gamma_3\gamma_4\gamma_5+\gamma_1\gamma_6}$$

3.2 Application of the Bradley-Terry Model on the Moves of a Game

The Bradley-Terry model can be applied to evaluate the possible moves of a given board position. Each legal move is considered as one team in competition with the other moves, and the members of the team are *features* specific to the game.

In the case of the game of Go, relevant features of a move are, e.g., the distance to the previous move, or the fact of being in immediate risk of capture by the opponent, a situation called *atari* by Go players. A γ value is assigned to each specific value of a feature. So, in the case of our example, we would have $\gamma(atari)$, $\gamma(not\ in\ atari)$, $\gamma(distance\ 2\ to\ the\ previous\ move)$, $\gamma(distance\ 3)$, $\gamma(distance\ 4)$, etc. A second commonly used feature of a move is the set of local patterns around the move. In this case, a γ value is assigned to each local pattern.

3.3 Learning the γ Values

When a relevant set of features is chosen, it is possible to learn the γ values automatically from game records. Let us assume that we know the move m_j that was chosen in n different board states of some game records (for example between strong players). For each board, the probability $P(m_j)$ that the move m_j was played can be written as a function of the γ values:

$$P(m_j) = \frac{\prod_{feature\ i\ \in\ m_j} \gamma_i}{\sum_{legal\ moves\ m}\left(\prod_{feature\ i\ \in\ m} \gamma_i\right)} \quad (2)$$

Then, we compute the γ values that maximize the following likelihood estimator: $L(\{\gamma_i\}) = \prod_{j=1}^{n} P(m_j)$.

In our current implementation, we use a variation of a gradient ascent method to find the likely maximum, but other methods have been proposed, in particular minorization-maximization algorithms [3].

The global efficiency of the learned Bradley-Terry model can be checked on a set of games between professional players, by evaluating if the moves played

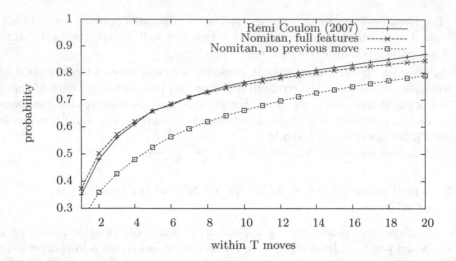

Fig. 1. Cumulative probability of finding the moves of the test set of game records in the top T static values

by the players are in the top T moves of the Bradley-Terry model. Figure 1 shows the cumulative probability of finding the move played in the game records in the top T static $P(m)$ values of the Bradley-Terry model. The cumulative probability with all the features of our program is close to the results by Coulom in 2007 [3]. We have not implemented "dynamic" features such as the *ownership* as described in [3]. This perhaps explains why Coulom's results are better for $T > 8$.

Moreover, we show the cumulative probability when the feature "distance to the previous move" of our program is disabled. We see that the results are significantly worse. It shows that this feature is useful, despite the fact that the Markov property of the game of Go implies that the static value of a move is theoretically independent from the last move played by the opponent.

3.4 Usage of the Bradley-Terry Model

The Bradley-Terry (BT) model is straightforwardly to use as a policy model of the Monte-Carlo simulations. For each legal move m of a given board, the BT model leads to a value $P(m) \in [0, 1]$, which reflects the probability that the move m would be played by a strong player. Then, a realistic simulation is obtained by selecting the move m with a probability $P(m)$.

In this paper, we focus on two other usages of the BT model, in the search tree of the UCT algorithm. We will describe (1) how to limit the number of searched moves, a method known as progressive widening, and (2) how to create an efficient static knowledge bias directly in the UCB formula.

The BT-model has even more possible applications, for example, to avoid unnatural moves when playing entertaining games against human players [14].

4   Progressive Widening

In this section we describe the progressive widening formula (Sect. 4.1) and the effect of progressive widening (Sect. 4.2).

4.1   Progressive Widening Formula

Compared to other classical board games, like draughts or chess, the number of legal moves on a typical Go board in the usual 19×19 size is very high, around 300 in the opening stage, and over 200 during most of the game. Searching all the legal moves would be unefficient, so we need to restrict the search to a small set of promising moves.

Coulom proposed the idea of *progressive widening* [3], which selects the top T moves with the biggest $P(m)$ value inside the Bradley-Terry model, and widens the search when the number of playouts increases. Chaslot et al. proposed independently the idea of *progressive unpruning* [4], which is essentially the same. This is efficient because we know from Fig. 1 that the probability of finding the best move within the top $T = 20$ moves of the Bradley-Terry model is as high as 84 %.

The above idea is used by many programs, such as THE MANY FACES OF GO [5] or AYA [13], but despite its widespread use, we have not found in the literature detailed data describing precisely the effect. We report here the results obtained with NOMITAN. We use a formula close to Coulom's formula and directly inspired by a formula by Hiroshi Yamashita, the developer of AYA [13]. With n the number of times a node has been visited, we write:

$$T = 1 + \frac{\log n}{\log \mu} \tag{3}$$

μ is a parameter that allows us to tune the formula. Figure 2 shows the number of searched moves given by the formula for typical values of the μ parameter.

4.2   Effect of Progressing Widening

Figure 3 shows the winning rate of our program against FUEGO on a 13×13 board. We have performed two experiments, one with the bias proposed in the next section ($C_{BT} = 0.6$, $K = 600$), and one without. We see that the μ parameter of progressive widening is clearly significant in both cases, and that the effect of progressive widening is relatively independent of the bias proposed in the next section.

When μ is too small, the winning rate drops because the search is distributed between too many candidate moves, and therefore the search is not sufficiently deep. On the contrary, when μ is too big, the search is focused on a too small number of candidate moves, and some important moves are not searched at all. It is interesting to note that the winning rate resists quite well to a very focused search (only 7 candidate moves are searched when $\mu = 5$), probably because the

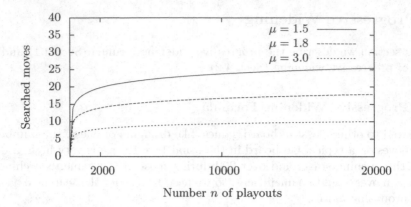

Fig. 2. Number of searched moves in function of the number of playouts, for different values of the μ parameter

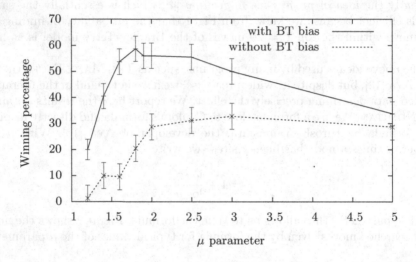

Fig. 3. Winning rate of our program vs Fuego in function of the μ parameter

small number of candidate moves is compensated by a deeper search that avoids local mistakes.

Lastly, we note that the best value of μ seems to be smaller when the bias of the next section is activated. This shows that it is better to search more moves, but only if the program has some way (other than only the plain UCB formula 1) to avoid the harmful scattering effect of a wider search.

5   Bradley-Terry Bias in the UCB Formula

We describe in this section how the Bradley-Terry model can be used as a bias in the UCB formula, and we study in detail the effect of some parameters with games against FUEGO and PACHI.

5.1   Bias in the UCB Formula

Adding a bias in the UCB formula is not a new idea. In fact, as far as we know, several strong Go programs such as ERICA [10] and ZEN [6] use some kind of bias in the UCB formula to enhance the selection of some preferred moves. This need can be explained easily by the fact that, in many cases, the winning ratio is almost the same between all the best candidate moves. In this kind of case, it is better to bias the choice of the move according to some policy learned offline instead of relying only on the small and noisy differences in the winning ratio.

Chaslot et al. proposed in 2009 to integrate the static knowledge with a bonus in the UCB formula [8]. They write formula 1 as follows:

$$\mu_j = \alpha \cdot \frac{w_j}{n_j} + \beta \cdot p_{rave}(m_j) + \left(\gamma + \frac{C}{\log(2 + n_j)} \right) \cdot H_{static}(m_j) \qquad (4)$$

where $p_{rave}(m_j)$ is the value of move m_j with the RAVE heuristic, and $H_{static}(m_j)$ some static value associated to move m_j according to a model learned offline. We are not sure if they also used the original exploration term of Eq. 1, or if it is completely replaced by the RAVE term.

The static knowledge model they used for H_{static} was partly hand-crafted and partly learned automatically for local patterns. They report that fine-tuning of the coefficients was needed to obtain good results, a difficulty possibly coming from the conflict between the RAVE heuristic term and the static term, both terms trying to bias the search in different directions during the startup phase.

In 2011, Shih-Chieh Huang proposed a related formula in his Ph.D. thesis [10]:

$$\mu_j = \alpha \cdot \left(\frac{w_j}{n_j} + C \cdot \sqrt{\frac{\ln n}{n_j}} \right) + (1 - \alpha) \cdot p_{rave}(m_j) + C_{PB} \cdot \frac{1}{n_j} \cdot H_{static}(m_j) \quad (5)$$

Huang does not detail how $H_{static}(m_j)$ is obtained, though we guess that it is based on the Bradley-Terry model.

In both [8] and [10], the amount of experimental data showing the effect of the bias is limited. One of our goals in this paper is to give more experimental evidence that such static bias is effective.

5.2   Bias Based on the Bradley-Terry Value

The bias that we propose here is inspired by formulas 4 and 5. It consists of adding a bonus term in the UCB formula to take directly into account the

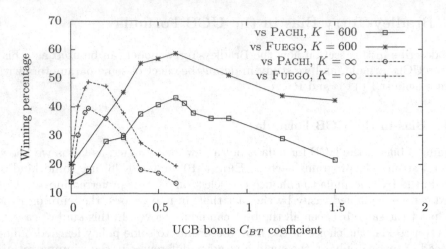

Fig. 4. Winning rate of NOMITAN in function of the UCB bonus C_{BT} coefficient

value $P(m_j)$ of the Bradley-Terry model described by Eq. 2. In a program that already contains a Bradley-Terry model for the simulation policy or for progressive widening, this is easy and immediate to implement. Formula 1 becomes:

$$\mu_j = \frac{w_j}{n_j} + C \cdot \sqrt{\frac{\ln n}{n_j}} + C_{BT} \cdot \sqrt{\frac{K}{n+K}} \cdot P(m_j) \qquad (6)$$

where C_{BT} is the main coefficient to tune the effect (small or strong) of this bias, and K is a parameter that allows us to tune the rate at which the effect decreases. After $3K$ visits of the parent node, the effect is 50 % of the initial value. A value of $K = \infty$ (in practice, it is sufficient to take a very large value) can be used to obtain a constant bias that does not decrease with the time.

5.3   Effect of the C_{BT} Coefficient

Figure 4 shows the effect of the main C_{BT} coefficient, against FUEGO, and against PACHI, with $K = \infty$ and $K = 600$. The progressive widening parameter is fixed to $\mu = 1.8$ in this experiment.

We can see that NOMITAN without the proposed bias is not so strong, with a winning rate of only 23 % against FUEGO and 14 % against PACHI. However, with the proposed bias, NOMITAN is of similar strength to these programs. With a constant bias ($K = \infty$), we reach 48 % against FUEGO and 39 % against PACHI, at $C_{BT} = 0.1$. The result is slightly better with a decreasing bias ($K = 600$), allowing us to reach 58 % against FUEGO and 43 % against PACHI, at $C_{BT} = 0.6$.

All curves display a characteristic mountain shape, which shows mainly that there is some ideal balance between the Bradley-Terry bias and the other UCB classical exploration and exploitation terms. In particular, if the bias is too big,

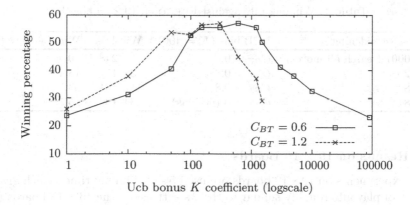

Fig. 5. Winning rate against FUEGO in function of the K parameter (logscale)

the program will select too frequently moves with a high static value, causing some local reading mistakes.

The peak for the decreasing bias is obtained at a bigger C_{BT} value than with a constant bias. This seems natural, since the stronger effect of the bias at the start of the search is compensated by its limited effect in the time.

5.4 Effect of the K Coefficient

Figure 5 shows the effect of the K parameter, with $C_{BT} = 0.6$ and $C_{BT} = 1.2$. The effect of the K parameter is robust, with a large range of values where the effect is stable, for example the range $[200, 2000]$ when $C_{BT} = 0.6$.

Figure 5 also shows that for a bigger bonus, a smaller value of K is better. It means that a bigger bonus needs to be decreased more quickly. In fact, the two curves look essentially the same, with an horizontal shift (in logscale).

The horizontal shift can be explained by a mathematical analysis of the bias in formula 6. When the number n of playouts increases sufficiently ($n \gg K$), the coefficient of the bias becomes close to $C_{BT} \cdot \sqrt{K} \cdot \frac{1}{\sqrt{n}}$.

This implies that we have somewhat similar effects for a constant value of $C_{BT} \cdot \sqrt{K}$. In function of $\log K$, we obtain that the curve for C_{BT2} is mainly an horizontal shift of the curve for C_{BT1} by $2 \log \frac{C_{BT2}}{C_{BT1}}$ units. This is what we obtain on Fig. 5, where the shift is $2 \log \frac{1.2}{0.6} = 0.6$ units (one unit is the distance between 1 and 10, or between 10 and 100).

6 Results on 19 × 19 Boards and Other Parameters

In this section, we report results on the 19×19 boards that confirm the results of the 13×13 boards (6.1). Moreover, we discuss the effect of some other parameters of NOMITAN, which are possibly related to the efficiency of the knowledge bias (6.2). Subsequently, we discuss the RAVE heuristic (6.3) and the features of the Bradley-Terry model (6.4).

Table 1. Winning rate against FUEGO on 19×19 boards

Progressive widening	UCB bias ($K = 400$)	Win-loss	Winning rate (%)
$\mu = 1.0001$ (search all moves)	$C_{BT} = 0.8$	0–248	0
$\mu = 1.2$	$C_{BT} = 0.8$	38–172	18.1
$\mu = 1.8$	$C_{BT} = 0.8$	117–101	53.7
$\mu = 1.8$	$C_{BT} = 0$ (no bias)	34–182	15.7

6.1   Results on 19×19 Boards

In the experiments on 19×19 boards, we used 5 s of thinking time, which gives a number of playouts roughly similar to the 2 s settings of the 13×13 boards (see Subsect. 2.2). The number of games for each data point is around 200 games, which leads to a 95 % confidence interval of ± 7 % on the winning rate, sufficient to say that the results are significant.

Table 1 shows the results of games against FUEGO with different settings. The results are quite similar to the 13×13 boards. The highest winning rate is obtained when both progressive widening and the knowledge bias are activated.

6.2   Final Selection of the Move

One of the effects of the knowledge bias in the Monte-Carlo Tree Search is an increase of the number of visits of moves with a high $P(m)$ value. It implies that the program should also rely on the number of visits of the moves, and not only on the winning ratio, when selecting the final move that will be played.

In NOMITAN, the final selection is a balance between the winning ratio and the number of visits, and we have increased the weight of the number of visits after introducing the static knowledge bias.

6.3   RAVE Heuristic

Contrary to most other strong Go programs, NOMITAN does not use the RAVE heuristic. We implemented a first trial, but without good results until now. It is possibly caused by the fact that many other parameters of the program have already been tuned without RAVE, making it difficult to find good parameters for RAVE.

It is not clear whether the knowledge bias described in this paper would be efficient or not in a program where RAVE is already tuned and gives good results. Investigating the relationship between such bias and the RAVE heuristic is one possible direction of future research.

6.4   Features of the Bradley-Terry Model

First, it must be noted that there is no need to use the same features in the Monte-Carlo simulations and in the knowledge bias. In particular, there is no need to use lightweight features in the knowledge bias.

Second, the features chosen for the Bradley-Terry model have a direct and important effect on the efficiency of the knowledge bias. For example, against Fuego, with $C_{BT} = 0.6$, $K = 600$ and $\mu = 1.8$, the winning rate of our program is 29 % without the "distance to the previous move" feature, and 58 % when used.

This property is quite important and shows that using better and richer features in the knowledge bias is a promising direction for improving the strength of Go programs.

7   Conclusion

We have described how static knowledge about the game of Go could be learned from game records with a Bradley-Terry model, and then used in a Go program in three different ways: (1) to improve the Monte-Carlo simulations, (2) to limit the number of searched moves (progressive widening), and (3) to bias the tree search.

The first usage of the Bradley-Terry model for improving the simulations is already well-known, so we have focused on the two other usages. In particular, we have described a straightforward and systematic way to include the static knowledge of the Bradley-Terry model as a bonus bias in the UCB formula.

We have performed a large number of games against two open-source Go programs, FUEGO and PACHI, allowing us to obtain clear experimental evidence that progressive widening and static knowledge bias are efficient. The systematic usage of the Bradley-Terry model in the three described ways allows our program to be competitive with FUEGO and PACHI.

References

1. Brugmann, B.: Monte Carlo Go. Technical report, Max-Planck Institute of Physics (1993)
2. Kocsis, L., Szepesvári, C.: Bandit based Monte-Carlo planning. In: Fürnkranz, J., Scheffer, T., Spiliopoulou, M. (eds.) ECML 2006. LNCS (LNAI), vol. 4212, pp. 282–293. Springer, Heidelberg (2006)
3. Coulom, R.: Computing Elo ratings of move patterns in the game of go. J. Int. Comput. Games Assoc. **30**(4), 198–208 (2007)
4. Chaslot, G., Winands, M., Uiterwijk, J., van den Herik, H., Bouzy, B.: Progressive strategies for Monte-Carlo tree search. New Math. Nat. Comput. **4**(3), 343–357 (2008)
5. Fotland, D.: Message on the Computer Go mailing list (2009). http://www.mail-archive.com/computer-go@computer-go.org/msg12628.html
6. Ojima, Y.: Message on the Computer Go mailing list (2009). http://www.mail-archive.com/computer-go@computer-go.org/msg10969.html
7. Enzenberger, M., Müller, M., Arneson, B., Segal, R.: Fuego - an open-source framework for board games and go engine based on Monte Carlo tree search. IEEE Trans. Comput. Intell. AI Games **2**(4), 259–270 (2010)

8. Chaslot, G., Fiter, C., Hoock, J.-B., Rimmel, A., Teytaud, O.: Adding expert knowledge and exploration in Monte-Carlo tree search. In: van den Herik, H.J., Spronck, P. (eds.) ACG 2009. LNCS, vol. 6048, pp. 1–13. Springer, Heidelberg (2010)
9. Gelly, S., Silver, D.: Monte-Carlo tree search and rapid action value estimation in computer go. Artif. Intell. **175**(11), 1856–1875 (2011)
10. Huang, S.C.: New heuristics for Monte Carlo tree search applied to the game of go. Ph.D. thesis, National Taiwan Normal University (2011)
11. Baudis, P.: MCTS with information sharing. Master thesis, Charles University in Prague (2011)
12. Baudiš, P., Gailly, J.: PACHI: state of the art open source go program. In: van den Herik, H.J., Plaat, A. (eds.) ACG 2011. LNCS, vol. 7168, pp. 24–38. Springer, Heidelberg (2012)
13. Yoshizoe, K., Yamashita, H.: Computer Go Theory and Practice of Monte-Carlo Method (2012, in Japanese). http://www.yss-aya.com/book2011/
14. Ikeda,K., Viennot, S.: Production of various strategies and position control for Monte-Carlo go - entertaining human players. In: IEEE Conference on Computational Intelligence in Games, pp. 145–152 (2013)

Investigating the Limits of Monte-Carlo Tree Search Methods in Computer Go

Shih-Chieh Huang[1] and Martin Müller[2]([✉])

[1] DeepMind Technologies, London, UK
[2] Department of Computing Science, University of Alberta, Edmonton, Canada
{shihchie,mmueller}@ualberta.ca

Abstract. Monte-Carlo Tree Search methods have led to huge progress in computer Go. Still, program performance is uneven - most current Go programs are much stronger in some aspects of the game, such as local fighting and positional evaluation, than in other aspects. Well known weaknesses of many programs include (1) the handling of several simultaneous fights, including the *two safe groups* problem, and (2) dealing with coexistence in seki.

After a brief review of MCTS techniques, three research questions regarding the behavior of MCTS-based Go programs in specific types of Go situations are formulated. Then, an extensive empirical study of ten leading Go programs investigates their performance in two specifically designed test sets containing *two safe group* and seki situations.

The results give a good indication of the state of the art in computer Go as of 2012/2013. They show that while a few of the very top programs can apparently solve most of these evaluation problems in their playouts already, these problems are difficult to solve by global search.

1 Introduction

In computer Go, it has been convincingly shown in practice [3] that the combination of simulations and tree search in Monte-Carlo Tree Search is much stronger than either simulation by itself, or other types of tree search which do not use simulations. It is also well known from practice that there remain several types of positions where most current MCTS-based programs do not work well. Examples are capturing races or semeai [9], positions with multiple cumulative evaluation errors [7], ko fights, and close endgames.

We present a number of case studies of specific situations in Go where many MCTS-based programs are known to have trouble. In each case, we develop a collection of carefully chosen test cases which illustrate the problem. We study the behavior of a number of state of the art Go programs on those test positions, and measure the success of their MCTS searches and their scaling behavior when increasing the number of simulations. Measures include the ability to select a correct move in critical situations and the estimated winning probability (the *UCT value*) returned from searches.

H.J. van den Herik et al. (Eds.): CG 2013, LNCS 8427, pp. 39–48, 2014.
DOI: 10.1007/978-3-319-09165-5_4, © Springer International Publishing Switzerland 2014

2   Four Classes and Three Research Questions

Three main components influencing the strength of current Go programs are (a) their simulation policy, (b) their tree search, and their (c) other in-tree knowledge, such as patterns. If we are trying to evaluate Go programs indirectly, without analyzing their source code, we can use test positions, which can be classified as follows.

1. *well-suited* positions: MCTS programs can solve them with either a good search, or with a good policy, or both.
2. *search-bound* positions: a stronger search can solve them, but simply using a stronger policy is not sufficient. Candidates might be hard tactical problems with unusual solution moves, such as filling your own eye.
3. *simulation-bound* positions: better simulation policies can play these positions well, but search alone will fail. Possible examples are seki and semeai.
4. *hard* positions (i.e., for current programs compared to humans): neither better policies nor (global) search helps. The main candidates are positions with multiple simultaneous fights [7], of which the combined depth is too deep to be resolved by global search, and which contain too many "surprising" moves to be "solved" in simulations.

Three related research questions are as follows.

1. Is it possible to design tests that separate the playout strength from the search strength of current Go programs?
2. Is it possible to identify types of positions where all current Go programs fail?
3. Is it possible to estimate the overall strength of programs from a dedicated analysis involving specific types of positions?

The current paper begins this investigation by analyzing two cases in detail, and showing test results for many of today's leading Go programs.

3   Two Test Scenarios

This section discusses and develops two test scenarios in detail: *Two safe groups* (TSG) and seki. All test data and experimental results are available at https:// sourceforge.net/apps/trac/fuego/wiki/MctsGoLimitsTestData.

3.1   Two Safe Groups

A main motivation for developing this test scenario is the observed poor performance of the FUEGO program when playing Black on a 9 × 9 board, especially with the large komi of 7.5 points [7]. A frequent 9 × 9 opening sees Black starting on the center point, then building a wall through the middle, while White establishes a group on each side. If both groups live on a reasonable scale, then it is hard for Black to obtain sufficient territory. Problems in FUEGO's play as

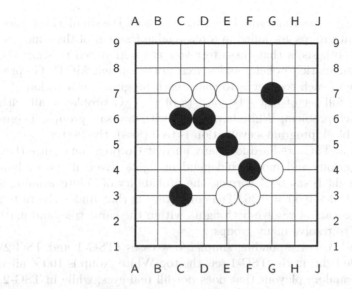

Fig. 1. Two safe white groups. Example from game Ping-Chiang Chou 4 Dan (W) - FUEGO-GB PROTOTYPE (B), Barcelona 2010. White wins by Black's resignation.

Black are often caused by over-optimism about being able to kill at least one of the white groups.

In this context, a *safe* group is defined to be a group that is safe with reasonable play, but may easily die during playouts. In contrast, a *solid* group already has a very definite eye shape, and will rarely die even in simulations. It has no weaknesses that could lead to accidental death during simulation. Figure 1 shows an example from a game between FUEGO and a professional player. Both white groups are safe, since it is easy to make two eyes for them with competent play. An amateur low-level Dan player should have no trouble winning with White. However, these groups are not solid at the level of FUEGO's simulations - it is easy for one or both of these groups to die in the course of a simulation following a not-too-informed policy. Accordingly, FUEGO's winning rate was close to 50 %, and remained so for the next thirty moves of this game, while for the professional human player the game was decided by move 12 at the latest.

Test Sets for One and Two Safe Groups. The tests are developed in sets of three positions (TSG, TSG-1, TSG-2) which are closely related. All positions are lost for Black. In the first position of the set (TSG), White has two groups which are *safe* but not *solid* according to the definition above. The second and third position in each set (TSG-1 and TSG-2) are very similar to the first one, but one of the two *safe* groups has been made *solid* by adding a few extra stones to simplify the eye space. Figures 2 and 3 show an example.

In creating the test sets and test scenarios, the authors attempted to control the difficulty, such that the game tree built by FUEGO's MCTS is sufficiently deep to (mostly) resolve the life and death for a single *safe* group, but not sufficiently deep for two or more groups. In informal tests with FUEGO, the simulations made enough mistakes that groups got killed in a reasonable fraction of the time.

Our working hypothesis is that these test sets are well suited to show the difference between simulation policies and search in the current MCTS Go programs. The purpose of each test set is to check each program's evaluation of a given position. The full set of test sets consists of 15 TSG problems, all with Black to play. In each problem, White is winning, with two safe groups. Therefore, the closer the black program's evaluation is to 0 (loss), the better.

Many test cases in TSG are hard for many current Go programs, since they involve two simultaneous and mostly independent fights. Even if the evaluation of each single fight is say 60 % correct, the probability of White winning a simulation is only $0.60 \times 0.60 = 0.36$. The conjecture is that under the tested conditions, programs cannot resolve both fights within the game tree, and must rely on simulations to resolve many groups.

In contrast, the two corresponding simpler test cases (TSG-1 and TSG-2) contain only a single fight: in the TSG-1 set, the top White group is 100 % alive for any reasonable random playout that does not fill real eyes, while in TSG-2, the bottom white group is 100 % alive.

3.2 Seki

Seki positions are designed to illustrate the "blindness" of global search when simulations are misleading. The goal was to create positions that are decided not by search, but by playouts. If crucial moves are missing, or are systematically misplayed in simulations, then the correct solution does not "trickle up" the tree until it is much too late.

The Seki Test Set consists of 33 cases, all with White to play. Figure 4 shows an example. At least one local seki situation is involved in each case. Furthermore, it is good to know that in each case, there is at least one correct answer that leads to White's win. Correct answers might include a pass move.

4 Experiments

Experiments were run with the TSG, TSG-1, TSG-2, and Seki Test Sets described above. Each test case is a 9×9 Go position. The aim is to measure a program's performance, including scaling, in these cases. Scaling is measured by increasing the number of playouts from 1 k to 128 k, doubling in each step.

For each test set, one or more regression test files in GTP format were created. The seki regression test checks a program's best move in a given test position. A program's answer is counted as correct only if its move is identical to one of the provided correct answers.

For the Two-Safe-Groups regression tests, in order to create problems with a simple, binary answer, first a threshold T in the range [0, 0.5] was selected. A program was requested to output -1 if its winning rate after search was at most T, and +1 otherwise. Since Black is losing in all test instances, the correct answer is -1 in all cases. Three tests with threshold T set to 0.3, 0.4 and 0.5 respectively were run.

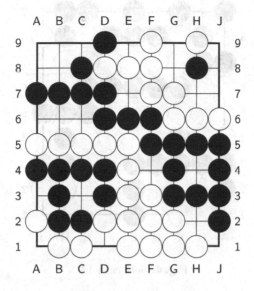

Fig. 2. TSG case 7.

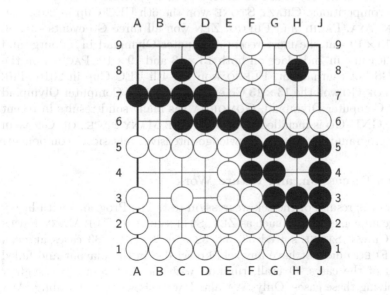

Fig. 3. TSG-1 case 7, with *solid* white group on bottom.

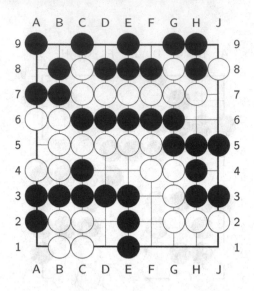

Fig. 4. Seki case 13.

4.1   The Programs

Table 1 lists the participating programs in alphabetical order, with the versions and processor specifications provided by their authors. These programs strongly represent the state-of-the-art of MCTS in view of their splendid records in recent computer Go competitions. CRAZY STONE won the 6th UEC Cup in 2013, followed by ZEN, AYA, PACHI and FUEGO. ZEN won all three Go events - 9 × 9, 13 × 13 and 19 × 19 - at the most recent Computer Olympiad in Tilburg, and STEENVRETER came in 2nd place on both 13 × 13 and 19 × 19. PACHI won the May 2012 KGS bot tournament. FUEGO won the 4th UEC Cup in 2010. THE MANY FACES OF GO won the 13 × 13 Go event at the 2010 Computer Olympiad and the 2008 Computer Olympiad. GOMORRA has many solid results in recent competitions. GNU GO is included along with THE MANY FACES OF GO as an example of a program with a large knowledge-intensive, "classical" component.

4.2   Results, Discussion, and Future Work

Figure 5 shows the results of the seki regression test. The Programs with heavy Go-knowledge implementations such as ZEN, STEENVRETER, THE MANY FACES OF GO and CRAZY STONE solved 33, 33, 32 and 30 out of 33 cases at peak respectively. FUEGO did not have any seki knowledge in the playout and failed in about half of the cases. Overall, running with more playouts just slightly helps with solving these cases. Only AYA and PACHI showed good scaling: AYA improved from 27 to 32 solved cases when scaling from 1 k to 128 k playouts, and PACHI improved from 21 to 26 solved. GNU GO solved 29 cases at level 10, which is not shown in the figure. The test strongly demonstrated that applying

Table 1. The programs participating in the experiments.

Name	Version	Processor Specification
AYA	7.36e	Core2Duo 1.83 GHz
CRAZY STONE	0013-07	Six-Core AMD Opteron(TM) 8439 SE
FUEGO	tilburg	Intel(R) Xeon(R) E5420 @ 2.50 GHz
GNU GO	3.9.1	Intel(R) Xeon(R) E5420 @ 2.50 GHz
GOMORRA	r1610	Intel(R) Core(TM) i7-860 @ 2.80 GHz
HAPPYGO	1.0	Intel(R) Xeon(R) E3-1225 @ 3.10 GHz
THE MANY FACES OF GO	0013-07	Intel(R) Core(TM) i7-3770 @ 3.40 GHz
PACHI	9.99	Intel(R) Xeon(R) E5420 @ 2.50 GHz
STEENVRETER	r123	Intel(R) Core(TM)2 Duo T9300 @ 2.50 GHz
STONEGRID	0.3.4 155	Intel(R) Core(TM) i7-2600 @ 3.40 GHz
ZEN	9.6	MacPro

seki knowledge to the playouts is crucial to MCTS programs to play in seki situations correctly.

Figures 6, 7, 8, 9, 10, 11 and 12 show the individual results of the two-safe-groups regression tests (TSG, TSG-1, TSG-2). Note that only 7 out of 11 programs supported querying their UCT value, and these results are shown.

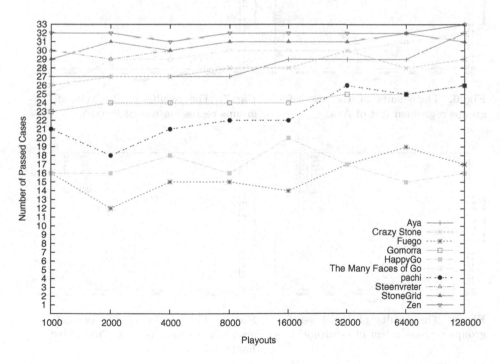

Fig. 5. The results of the seki regression test.

TSG is a difficult test for these MCTS programs because most of them failed in more than half of the cases, even for $T = 0.5$. FUEGO did not solve any case for $T = 0.3$, even with 128 k playouts. GOMORRA solved only one case, with 8 k playouts. This indicates that many MCTS Go-playing programs cannot evaluate life-and-death and semeai situations correctly without specific knowledge. Programs solved more cases for $T = 0.4$ and $T = 0.5$, indicating that their evaluations of these positions are mostly between 0.3 and 0.5, far from the optimal value 0. The TSG problem, the difference in performance between TSG on one side and TSG-1 and TSG-2 on the other side, is most pronounced in FUEGO, GOMORRA, PACHI, THE MANY FACES OF GO and STEENVRETER. It is less pronounced in AYA, which surprisingly does relatively poorly even with one solid group. TSG-1 and TSG-2 can often be resolved by search, since there is only a single fight. STEENVRETER solved most of the cases of TSG-1 and TSG-2 and did well for TSG with a $T = 0.5$ cutoff. ZEN did by far the best overall, and even solved most cases of TSG. While the techniques used in ZEN have not been

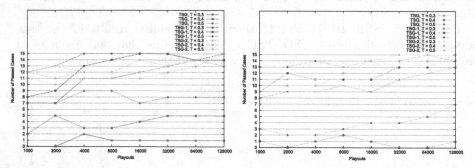

Fig. 6. The results of the two-safe-groups regression test of AYA.

Fig. 7. The results of the two-safe-groups regression test of FUEGO.

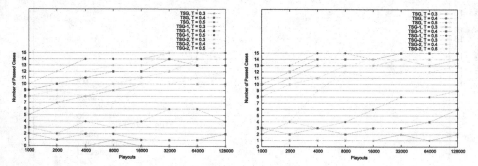

Fig. 8. The results of the two-safe-groups regression test of GOMORRA.

Fig. 9. The results of the two-safe-groups regression test of THE MANY FACES OF GO.

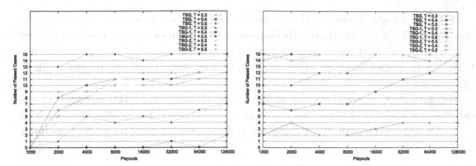

Fig. 10. The results of the two-safe-groups regression test of PACHI.

Fig. 11. The results of the two-safe-groups regression test of STEEN-VRETER.

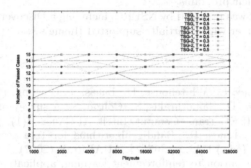

Fig. 12. The results of the two-safe-groups regression test of ZEN.

published, its authors have publicly described ZEN as using knowledge-heavy, slow but very well-informed playouts.

We believe that this work provides a valuable first step towards analyzing the state of the art in computer Go. One lesson is that programs which are quite similar in strength can exhibit quite different behavior in terms of their simulation policies and search scaling. We hope that this approach can be helpful for current and future Go programmers to analyze the behavior of their programs. Future work includes: (1) developing more test sets, such as semeai, more than two groups, connection problems, and endgames; (2) scaling to more simulations; and (3) developing similar tests for other games such as Hex.

5   Related Work

There is a large and quickly growing literature about applications of MCTS to games and many other domains. See [1] for a recent survey. Much of that work focuses on algorithm variations, parameter tuning etc. There is much less work on identifying limitations.

One well-known result is the tower-of-exponentials worst-case convergence time for UCT [2,6]. Work by Ramanujan and Selman shows that UCT can be over-selective and miss narrow tactical traps in chess [8]. In practice, use of MCTS in games such as chess and shogi has been confined to small niche applications so far.

Negative experimental results on the correlation between the strength of the policy as a stand-alone player vs. its strength when used for guiding simulations in MCTS were shown in [4]. Follow-up work on simulation balancing includes [5,10].

Acknowledgements. This project would not have been possible without the support of all the Go program's authors. Many of them supported us by implementing extra GTP commands in their programs, and by helping to debug the test set through testing early versions with their programs.

Financial support was provided by NSERC, including a Discovery Accelerator Supplement grant for Müller which partially supported Huang's stay.

References

1. Browne, C., Powley, E., Whitehouse, D., Lucas, S., Cowling, P., Rohlfshagen, P., Tavener, S., Perez, D., Samothrakis, S., Colton, S.: A survey of Monte Carlo tree search methods. IEEE Trans. Comput. Intell. AI Games 4(1), 1–43 (2012)
2. Coquelin, P.-A., Munos, R.: Bandit algorithms for tree search. In: Parr, R., van der Gaag, L. (eds.) UAI, pp. 67–74. AUAI Press (2007)
3. Gelly, S.: A contribution to reinforcement learning; application to computer-Go. Ph.D. thesis, Université Paris-Sud (2007)
4. Gelly, S., Silver, D.: Combining online and offline knowledge in UCT. In: ICML '07: Proceedings of the 24th International Conference on Machine Learning, pp. 273–280. ACM (2007)
5. Huang, S.-C., Coulom, R., Lin, S.-S.: Monte-Carlo simulation balancing in practice. In: van den Herik, H.J., Iida, H., Plaat, A. (eds.) CG 2010. LNCS, vol. 6515, pp. 81–92. Springer, Heidelberg (2011)
6. Kocsis, L., Szepesvári, C.: Bandit based Monte-Carlo planning. In: Fürnkranz, J., Scheffer, T., Spiliopoulou, M. (eds.) ECML 2006. LNCS (LNAI), vol. 4212, pp. 282–293. Springer, Heidelberg (2006)
7. Müller, M.: Fuego-GB Prototype at the human machine competition in Barcelona 2010: a tournament report and analysis. Technical report TR 10–08, Dept. of Computing Science. University of Alberta, Edmonton, Alberta, Canada (2010). https://www.cs.ualberta.ca/research/theses-publications/technical-reports/2010/TR10-08
8. Ramanujan, R., Selman, B.: Trade-offs in sampling-based adversarial planning. In: Bacchus, F., Domshlak, C., Edelkamp, S., Helmert, M. (eds.) ICAPS, pp. 202–209. AAAI (2011)
9. Rimmel, A., Teytaud, O., Lee, C.-S., Yen, S.-J., Wang, M.-H., Tsai, S.-R.: Current frontiers in computer Go. IEEE Trans. Comput. Intell. AI Games 2(4), 229–238 (2010). (Special issue on Monte-Carlo Techniques and Computer Go)
10. Silver, D., Tesauro, G.: Monte-Carlo simulation balancing. In: Danyluk, A., Bottou, L., Littman, M. (eds.) ICML. ACM International Conference Proceeding Series. vol. 382, pp. 945–952. ACM (2009)

Programming Breakthrough

Richard Lorentz$^{(\boxtimes)}$ and Therese Horey

Department of Computer Science, California State University,
Northridge, CA 91330-8281, USA
lorentz@csun.edu, therese.horey57@my.csun.edu

Abstract. Breakthrough is an abstract strategy board game that requires considerable strategy in the early stages of the game but can suddenly and unexpectedly turn tactical. Further, the strategic elements can be extremely subtle and the tactics can be quite deep, involving sequences of 20 or more moves. We are developing an MCTS-based program to play Breakthrough and demonstrate that this approach, with proper adjustments, produces a program that can beat most human players.

Keywords: MCTS · Breakthrough · Evaluation function

1 Introduction

Breakthrough is a fairly new game, invented in 2000 by Dan Troyka and was the winner of the 2001 about.com 8 × 8 Game Design Competition [12]. It is played on an 8 × 8 board. The first two rows are populated with White's pieces and the last two rows with Black's. The first row is White's home row and Black's goal row. The eighth row is Black's home and White's goal. White moves his[1] pieces towards his goal, one square forward at a time, either straight or diagonally. White may not move to a square already occupied by a White piece and may only capture a Black piece if it is a diagonal move. White cannot move straight forward to a square occupied by a Black piece. Black moves similarly in the other direction. The first player to reach his goal row or capture all the opponent's pieces is the winner. Note that a draw is impossible in Breakthrough.

Figure 1 shows two Breakthrough positions, both with White to move.[2] The one on the left is the starting position and the one on the right is after 22 moves have been played. From the second position the White piece on g2 may move either to g3 or h3, the piece on f3 may either move to g4 or capture on e4, and the piece on f4 may make either diagonal move but cannot move forward since it is blocked by Black.

Figure 1 also gives a sense of the highly strategic nature at the early stages of the game. From the starting position certainly no plans for breaking through and

---

[1] For brevity, we use 'he' and 'his' whenever 'he or she' and 'his or her' are meant.
[2] All Breakthrough figures are taken from the Little Golem website [13].

H.J. van den Herik et al. (Eds.): CG 2013, LNCS 8427, pp. 49–59, 2014.
DOI: 10.1007/978-3-319-09165-5_5, © Springer International Publishing Switzerland 2014

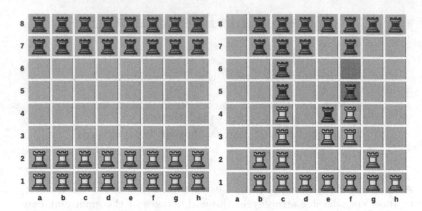

Fig. 1. Sample Breakthrough positions.

reaching the goal are evident. Even more than 20 moves into the game no obvious plans emerge for breaking through the opponent and reaching the goal. Instead, one must try to position one's pieces in such a way as to be ready to make a tactical break should one become available while simultaneously defending against a similar attack from the opponent. Speaking with experts we have learned that three strategic goals are as follows.

1. Keep pieces on the home row for as long as possible, especially the b, c, f, and g columns since those four can effectively defend an attack by a single opponent's piece.
2. Do not give up tempi by exchanging pieces in enemy territory.
3. Keep an even distribution of pieces between squares with even parity and those with odd parity.

To get a feeling for the tactical nature of the game consider the two positions in Fig. 2. In the left diagram Black has just moved 54. h5-h4 and appears to be making threats to break through on the right side – which can be realized if another piece is brought to g4. However, it turns out that if White plays 55. d5-e6 he can now force a win on the left side of the board. One likely continuation is 56. d8-d7 57. e6xd7 58. c8xd7 59. c4-c5 leading to the position on the right where we can now see that after Black exchanges twice on c5 White can break through by advancing to b6 (or d6 if Black plays a8-a7) and win the game on move 69, which is a 16-ply tactical sequence. This game is slightly longer than average in that the average length of a game between strong players is in the low 60's.

The work reported on here shows that a Monte-Carlo Tree Search (MCTS) programming approach to Breakthrough can simultaneously deal with both the strategic and the tactical aspects of the game. Most of the analysis is done using self-tests but it is, of course, necessary to have external validation of our results. Since neither author considers themselves an expert Breakthrough player, tests against us would be unreliable. Instead, we have elected to compete our program

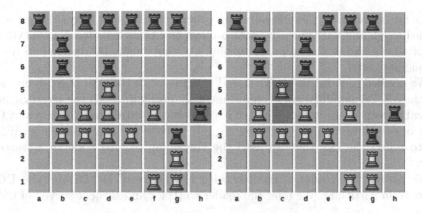

Fig. 2. A tactical situation in Breakthrough.

on the turn-based[3] Little Golem game server [13]. Allowing human opponents large amounts of time to select their moves is in many ways the ultimate challenge for game-playing programs since it is well known that longer time controls favor the human.

The rest of this paper is organized as follows. In Sect. 2 we outline how the current structure of our Breakthrough program evolved. We also explain some of our high-level decisions as applied to MCTS. In Sect. 3 we provide details of the different components and provide test results showing the effectiveness of various algorithmic choices. Finally, in Sect. 4 we summarize our results and discuss future work.

2 Wanderer, a Program that Plays Breakthrough

We were first attracted to this game because of its interesting mix of strategy and tactics as described in the previous section. Also, the state space is quite large given that the game is played on an 8 × 8 board with the possibility of 32 pieces being in play at the same time. Finally, the branching factor is not small in that a typical position will allow 20 or more legal moves. Considering all of these factors it was not at all clear how strong of a program we could expect to construct. We note that others have also found Breakthrough to be an interesting research domain [7,9].

Both authors are new to this game so we needed access to other Breakthrough players both as potential sparring partners for our program and to provide insight into the strengths and weaknesses of the program as it evolved. We decided to launch our program on the Little Golem (LG) web site [13] where there is a fairly large Breakthrough playing community. This choice also added to our challenge since long thinking times favor human competitors.

---

[3] A turn-based server allows players days to make moves rather than minutes as is the case in over-the-board play.

As far as we know there is only one other program that plays the game of Breakthrough and, conveniently, it also plays on LG. It is called NATALIABOT, is alpha-beta based, and appears to play at what might be referred to as an intermediate level.

We decided to write an MCTS-based program with the goal of studying how well the MCTS paradigm could simultaneously deal with both the strategic and tactical aspects of the game. We gave it the name WANDERER, which is also the name of our Havannah playing program [6], to facilitate the interface with LG and to reflect the fact that we used the Havannah playing code as a starting point.

We assume the reader is familiar with the basics of MCTS and the UCT approach in particular. We refer the reader to [1] for an excellent survey of these topics.

2.1 Basic Structure

We began by building a basic Monte-Carlo program. The major design decision at this point revolved around finding the best way to generate fast random playouts. We found it most efficient to focus on the location of the pieces. That is, rather than scanning the board for legal moves, we maintain the pieces in a separate data structure and then generate legal moves for each piece. Fortuitously, this also aided in the efficiency of the evaluation function that was incorporated into the program later and is discussed below.

Adding UCT-based MCTS was straightforward. We also incorporated some standard improvements (see [6], for example), namely, a transposition table, a solver for the endgame, improved random playouts, and initialized win and visit counts in the MCTS tree in an effort to introduce some specific domain knowledge. Details and test results are outlined in the next section.

Next we tuned parameters. Specifically, we experimented with different UCT constants, adjusted the various initial win and visit counts, and tweaked the probabilities that encouraged certain moves in the random playouts. Based on self-tests we felt we had a reasonably stable program and so we started playing WANDERER on LG. However, we did not get the kinds of results we had hoped for. The solver proved quite useful in the late endgame where it was often able to find forced wins before its human opponent. Yet, at other stages of the game it tended to founder. It frequently violated all of the strategic concepts mentioned in the introduction and worse, it did not seem to have any understanding of the value of keeping its own pieces. It was more than willing to sacrifice pieces, even in the early stages of the game, for no apparent gain. With so many pieces on the board it was unable to discern any difference between positions where the number of pieces it had was the same as its opponent and positions where it had one or two fewer pieces. But, having an extra piece early in the game (before any real tactics emerge) usually proves to be a significant advantage. It seems that the random playouts were simply not providing sufficient information and our attempts to add knowledge to the playouts always met with failure. Playing only a very few games on LG quickly revealed these flaws. Even the authors were able to defeat the program. Overall it was playing at what might generously be called an intermediate level, weaker than NATALIABOT.

2.2 The Evaluation Function

In the past, we have experimented with using evaluation functions in an MCTS setting. Whereas we had no success with our Havannah playing program [6], we had considerable success with INVADER, our Amazons playing program [5]. The difference between these two cases is, of course, that it is possible to find a reasonably accurate static evaluation function for Amazons while that does not seem to be the case for Havannah. Kloetzer et al. had similar success with their Amazons program [4], as did Winands and Björnsson with an LOA program [10], and historically Sheppard did something similar with a Scrabble® playing program in a pure MC setting [8].

We felt that there were a number of features in a Breakthrough position that would be easy to quantify in an evaluation function. Certainly the most obvious, and the one that originally motivated our research into evaluation functions, is the relative number of pieces. Other features such as distance to the goal, open lines to the goal, distribution of pieces on the board, strength of a piece, etc., should all be quantifiable in an evaluation.

The remaining question is, how best to use the evaluation? Two obvious choices are: (1) to provide guidance when initializing the win and visit counts in the MCTS tree and (2) to help skew the random playouts in the direction of more reasonable moves. Similar to our experiences with INVADER we had little success in either area. Instead, as was the case with INVADER, our success came with terminating the playouts early and using the evaluation function to report a win or a loss based on whether the value is positive or negative. With this enhancement WANDERER was able to play at a significantly higher level. It now easily beats the authors and though results come very slowly from the turn-based server LG, it seems now to be a rather strong player. We will provide more details in Sect. 4 below.

3 Internal Workings

To better understand some of the improvements made to the basic MCTS algorithm we define five terms explicitly below. Consider the two positions (a middle game position and an endgame position from the same game) in Fig. 3.

We call a piece *safe* if it cannot be immediately captured in a sequence of capturing moves, that is, the number of enemy pieces attacking that piece is no more than the number of friendly defenders. A move is safe if the piece moved ends up to be safe. All pieces in the left diagram of Fig. 3 are safe. However if White were to move f3-g4 it would no longer be safe as it could be captured immediately. Similarly, moving e4-d5 is also not safe since Black will come out one piece ahead after the consecutive captures on the d5 square. The black move d6-d5 is safe but e6-d5 is not.

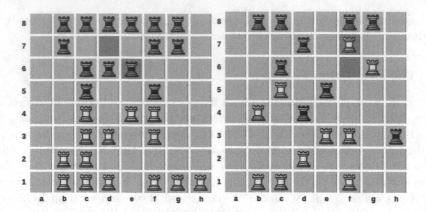

Fig. 3. Middle game and endgame position.

In the diagram on the right, if it is White's turn to move White has the winning move f7-e8. We call this a *winning move*. Black can prevent this immediate win with the capture g8xf7. This is a *saving move*. We extend these ideas back one move. If White did not have a piece on f7, then Black's h3-h2 guarantees a win on the next move and is called a *win threat*. If White had a piece on g2 then White could parry this win threat by capturing and we call this a *parry* as it prevents the win.

We now discuss in more detail the various improvements to MCTS that we implemented. The test results reported for the full MCTS version with evaluation function were from 500 game matches where each player was allotted 10 s per move on an Intel® Xeon® CPU E5-2680 @ 2.70 GHz. The tests performed on versions not involving the evaluation function were 200 game matches.

3.1 Improved Playouts

It is well known that improving the random playouts can significantly improve an MCTS program. As an early example, see [3]. It is also known that properly improving the playouts is a bit of a black art. Having the playouts play better or more natural looking moves does not guarantee improved performance. The quality of the game moves made can actually go down even when the playouts look better.

We began by doing some basic experiments with a purely MC version of WANDERER. A natural thought is that during a playout if a winning move is available, make it. Certainly waiting to make such a move will only introduce noise in the playout. Surprisingly, WANDERER so enhanced only won 40 % of its games against the purely random MC version. But if we add a balancing factor that a playout will also always make a saving move if one is available then it won 85 % of its games. We have no satisfactory explanation for this counterintuitive behavior.

We performed similar tests with the full MCTS version before the evaluation function was added. The results were similar to the MC results, but not quite so stark. In this case adding winning moves gave a win rate of 56 %, adding saving moves without winning moves gave a win rate of 81 %, and adding both produced a winning rate of 92 %. These two enhancements apparently complement each other very well.

Adding the evaluation function changes the complexion of WANDERER quite a bit. As described below, only 4 moves towards a random playout are made before the evaluation is invoked so it is not clear how much can be accomplished by improving the playouts. Still, by adding winning, saving, threat, and parry moves to the playout and giving safe moves triple the chance of being made over non-safe moves we find that WANDERER wins 61 % of the games against a version without these improvements.

3.2 Transposition Tables

A certain number of transpositions can be expected to arise in the MCTS tree so it is natural to implement hash tables to deal with move transpositions. Rather than have multiple nodes representing the same position we can point to a single copy of the node, turning our tree into a DAG. We use standard 64-bit Zobrist hashing [11]. We store the full hash value in the hash table (the size of the hash table is much smaller than the number of different hash values, of course), not the entire position. In the case of a collision we attempt one rehash and if a second collision occurs no hashing is done. In the interest of speed no replacement policy is implemented. This means that the first value placed in the hash table will remain in the hash table. Also, in the (very) unlikely event two positions have the same hash value we are forced to live with the fact that they will be considered the same for hashing purposes.

Our tests show that hashing does provide some benefit. A version of WANDERER with hashing wins 57 % of its games versus a version without. We have observed some further advantage when we have increased the size of the hash table and so this introduces an interesting optimization problem we have yet to tackle, namely, how to balance the high memory demands of the MCTS tree with the size of the hash table.

3.3 A Solver

It is known that incorporating a solver into an MCTS tree is fairly straightforward and can provide significant benefit [6,10]. We introduce a solver at two different levels. First, we add a flag to the MCTS nodes so that when a win or loss is detected in the tree, the node can be so flagged and the value can be backed up the tree in standard fashion.

Secondly, we do some advanced checking when the MCTS algorithm expands a node. Before we actually add a node to the tree we do some static evaluating to see if the position corresponding to the node allows a forced win or loss. If it is a win we flag it as such and then back up the value as appropriate. If it is a loss, we do not create the node (since we do not want to make such a move) and so nothing is added to the tree.

The actual checking is done incrementally as follows, where each subsequent step is carried out only if none of the previous steps apply. If the move creates a winning configuration the node is flagged as a win. If the opponent can make a winning move, the node is flagged as a loss. If the move is a win threat (it cannot be captured and it is one row from the goal) then the node is flagged as a win. Similarly (ignoring implementation details) if the opponent has a win threat, flag the node as a loss. If none of these apply, do one final check to see if it is possible to move to a square 2 rows from the goal in such a way that it cannot be captured. If so, flag the win.

We were a bit surprised to find that doing these checks did little, if anything, to improve the strength of WANDERER. It seems that simply flagging wins when found in the tree is algorithmically sufficient to detect correctly and quickly forced wins. However, by reducing the number of nodes in the MCTS tree, they did significantly reduce memory requirements. Since our goal is to produce a program that can play well in a turn-based setting, saving memory was useful in that it allowed WANDERER to think longer before exhausting available memory.

Even though the solver is able to detect wins from quite a distance, sometimes greater than 20 moves, its influence on the program is only modest. This is probably because WANDERER will always make reasonable moves even if it does not happen to know that it has a forced win. Still, WANDERER playing against a handicapped version that does not have the solver wins 60 % of the games.

3.4   Initializing Win Values in the MCTS Tree

When creating new nodes for the MCTS tree, the win counts and the total playout counts are usually initialized to zero. However, if there is prior knowledge of the quality of the move represented by a node relative to its siblings, then it sometimes makes sense to initialize these values to something other than zero that reflects the perceived relative value of that position [2]. We do this by always initializing the number of visits in a new node to be 100 and then setting the number of wins according to how much we like the move.

Our original expectation was we would be able to use the evaluation function to help us judge the values of various moves and set initial win counts accordingly. For reasons not yet clear to us, we were unable to get satisfactory results with this approach. We are confident that the evaluation function is not to blame since it proved quite useful when used with early playout termination (see below). One possibility is that since the evaluation is somewhat slow, the time penalty was just too great.

Nevertheless, we were able to obtain good results with this technique. We based the initial win values on two factors: (1) how close the move is to the goal and (2) whether or not it is a safe move. If a move is safe then we set the initial win count to 100, 95, 85, 75, or 60 according to whether the move is on the goal, 1 row away from the goal, etc. Also, if a move is not safe but is a capture we initialize the win count to 60. In all other cases it is set to 30. With this simple arrangement the program wins 89 % of its games versus a version that does not have this feature.

Initializing win counts actually benefits our program in two ways. It biases good moves with few visits in the MCTS tree so that we do not have to wait for the random playouts to sort them out. This biasing also significantly aids the solver. Even in 10 s games we see that forcing sequences tend to be found on average two moves sooner when this initialization is in place.

3.5   An Evaluation Function

One of the advantages of MCTS is that strong game-playing programs can be written without the need of evaluation functions. Go and Havannah are examples of games where no good evaluation functions are known, yet strong MCTS-based programs exist. Still, just because a reliable evaluation function exists, it does not automatically mean that the best programs will be alpha-beta based. Amazons provides a good example. Even though many good alpha-beta based programs existed for many years, creating an MCTS program and exploiting an existing evaluation function produced the world's strongest Amazons playing program [5]. We believe that using evaluation functions in MCTS programs is an under-appreciated and underused technique.

Our evaluation function for Breakthrough is quite simple. There are just four components. The first is piece count. The side with more pieces on the board is given 10 points for each extra piece.

Second, we evaluate where the pieces are on the board. This part of the evaluation is based on our own observations supplemented by advice from expert players. Pieces closer to the goal are worth more than pieces closer to home. Pieces on the edge are not worth quite as much. It is not a good idea to move pieces off the home row too early (otherwise it is easier for the enemy to break through), so pieces on the home row are worth more than pieces further up the next few rows. Finally, when moving pieces off the home row, pieces on the b, c, f, and g columns should be moved last. This is because those four columns alone provide sufficient protection against a single invader and so the enemy will need to penetrate with two pieces in order to reach the goal. We implement this idea very simply with an array that shows the value of a piece according to its location on the board. Figure 4 shows this array where the top row is the home row of the player.

The third element of the evaluation is to reward safe pieces. We do this by giving them a 50 % bonus over their array value.

The fourth component of the evaluation function attempts to estimate the likelihood of a piece breaking through. We do this by counting the number of empty squares or squares occupied by friendly pieces in the 2 by 3 rectangle in front of the piece. We give a further bonus to that piece according to how many such squares there are. When competing a version of the program that uses this last idea against one that does not, the one that uses it wins 53 % of the games.

As in our Amazons program we have so far been unsuccessful in using the evaluation for improving the random playouts. Whereas we have had some

```
 5, 15, 15,  5,  5, 15, 15,  5
 2,  3,  3,  3,  3,  3,  3,  2
 4,  6,  6,  6,  6,  6,  6,  4
 7, 10, 10, 10, 10, 10, 10,  7
11, 15, 15, 15, 15, 15, 15, 11
16, 21, 21, 21, 21, 21, 21, 16
20, 28, 28, 28, 28, 28, 28, 20
36, 36, 36, 36, 36, 36, 36, 36
```

Fig. 4. Values of pieces according to their location on the board.

success in Amazons using the evaluation to set initial win values, we have not been able to match such successes here. However, using the evaluation has been rather successful when used for cutting the playouts short, both here and in Amazons. With some tuning we find that applying the evaluation after only 4 random moves seems best and wins approximately 90 % of the games versus a version without this feature. We tested a version that cuts off after 5 random moves (the value we originally thought was best) against versions that cut off after 2, 3, 4, 6, and 7 moves and found that it won 57 %, 51 %, 46 %, 57 %, and 64 % of its games, respectively.

4   Results and Remarks

We have shown that MCTS can be successfully used for programming Break-through. In summary, the evaluation function is responsible for capturing the strategic ideas while the MCTS tree and the solver combine to provide some strong tactical skills. We have had good results using evaluation functions both in Breakthrough and in Amazons and suggest that using evaluation functions in MCTS is under-appreciated.

Since our stated goal is to perform well on LG, a turn-based server, results come slowly. We allow WANDERER a maximum of ten minutes to consider its move (less if it runs out of memory) but an opponent may take as long as 36 s. Because of this slow pace an opponent may have actually faced up to a half a dozen different versions of WANDERER. However, the current version described here has been playing unchanged for some time now and some trends are beginning to emerge.

Strong opponents are able to build imposing positions and attack weak points that WANDERER is still oblivious to. One successful ploy used by opponents is to develop very slowly, find and occupy holes in WANDERER's position, wait for WANDERER to vacate some squares in its home row, and then finally break through towards those empty squares. Such losses seem to show weaknesses in both tactics and the strategy. Strategically, WANDERER does not show sufficient patience and tactically it sometimes does not see the threats to its home row.

Nevertheless, with tuning and tweaking WANDERER continues to improve. It has beaten a number of the top players a number of times. Weaknesses against WANDERER are getting harder to find and exploit.

LG provides a rating system for its games. The average Breakthrough rating is approximately 1500, the top rated player is 2253, and the tenth best is 1913. We observed that the version without an evaluation function displayed many flaws. Based on the few games played we estimate the rating of that version to be in the 1600's. As of this writing the current version has a rating of 1910, the twelfth highest on LG. We anticipate it can make it and stay in the top 10, but whether it can make it to the top 2 or 3 remains to be seen. Because of the pace of the games on LG data comes very slowly so only time will tell.

References

1. Browne, C., Powley, D., Whitehouse, D., Lucas, S., Cowling, P., Rohlfshagen, P., Tavener, S., Perez, D., Samothrakis, S., Colton, C.: A survey of Monte Carlo tree search methods. IEEE Trans. Comput. Intell. AI Games 4(1), 1–49 (2012)
2. Gelly, S., Silver, D.: Combining online and offline knowledge in UCT. In: Proceedings of the 24th International Conference on Machine Learning, ICML 07, pp. 273–280. ACM Press, New York (2007)
3. Gelly, S., Wang, Y., Munos, R., Teytaud, O.: Modification of UCT with patterns in Monte-Carlo go. Technical report 6062, INRIA, France (2006)
4. Kloetzer, J., Iida, H.: Bouzy, B: The Monte-Carlo approach in Amazons. In: Computer Games Workshop, Amsterdam, The Netherlands, pp. 113–124 (2007)
5. Lorentz, R.J.: Amazons discover Monte-Carlo. In: van den Herik, H.J., Xu, X., Ma, Z., Winands, M.H.M. (eds.) CG 2008. LNCS, vol. 5131, pp. 13–24. Springer, Heidelberg (2008)
6. Lorentz, R.: Experiments with Monte-Carlo tree search in the game of Havannah. ICGA J. 34(3), 140–150 (2011)
7. Saffidine, A., Jouandeau, N., Cazenave, T.: Solving BREAKTHROUGH with race patterns and job-level proof number search. In: van den Herik, H.J., Plaat, A. (eds.) ACG 2011. LNCS, vol. 7168, pp. 196–207. Springer, Heidelberg (2012)
8. Sheppard, B.: Towards perfect play of scrabble. Ph.D. thesis, IKAT/Computer Science Department, University of Maastricht (2002)
9. Skowronski, P., Björnsson, Y., Winands, M.: Automated discovery of search-extension features. In: Proceedings of the 13th International Advances in Computer Games Conference (ACG2011), Tilburg, The Netherlands, pp. 182–194 (2011)
10. Winands, M.H.M., Björnsson, Y.: Evaluation function based Monte-Carlo LOA. In: van den Herik, H.J., Spronck, P. (eds.) ACG 2009. LNCS, vol. 6048, pp. 33–44. Springer, Heidelberg (2010)
11. Zobrist, A. L.: A new hashing method with application for game playing. Technical report 88, Computer Sciences Department, University of Wisconsin, Madison, Wisconsin, (1969) Reproduced in ICCA J. 13(2), 69–73 (1990)
12. http://boardgames.about.com/od/freeboardcardgames/a/design_winners.htm
13. http://www.littlegolem.net/jsp/index.jsp

MoHex 2.0: A Pattern-Based MCTS Hex Player

Shih-Chieh Huang[1,2], Broderick Arneson[2], Ryan B. Hayward[2(✉)],
Martin Müller[2], and Jakub Pawlewicz[3]

[1] DeepMind Technologies, London, UK
[2] Computing Science, University of Alberta, Edmonton, Canada
hayward@ualberta.ca
[3] Institute of Informatics, University of Warsaw, Warsaw, Poland

Abstract. In recent years the Monte Carlo tree search revolution has
spread from computer Go to many areas, including computer Hex. MCTS-
based Hex players now outperform traditional knowledge-based alpha-
beta search players, and the reigning Computer Olympiad Hex gold
medallist is the MCTS player MoHex. In this paper we show how
to strengthen MoHex, and observe that—as in computer Go—using
learned patterns in priors and replacing a hand-crafted simulation policy by a softmax policy that uses learned patterns significantly increases
playing strength. The result is MoHex 2.0, about 250 Elo points stronger
than MoHex on the 11×11 board, and 300 Elo points stronger on the
13×13 board.

1 Introduction

In the 1940s Piet Hein [22] and independently John Nash [26–28] invented Hex,
the classic two-player alternate-turn connection game. The game is easy to implement — in the 1950 s Claude Shannon and E.F. Moore built an analogue Hex
player based on electrical circuits [29] — but difficult to master. It has often
been used as a testbed for artificial intelligence research.

Around 2006 Monte Carlo tree search appeared in Go [11] and soon spread
to other domains. The four newest Olympiad Hex competitors — MoHex from
2008 [4], YOPT from 2009 [3], MIMHEX from 2010 [5], PANORAMEX from 2011
[20] — all use MCTS.

In this paper we show how to strengthen MoHex, the reigning Computer
Olympiad Hex gold medallist [20]. Among several improvements, we observe
that — in Hex as in Go — replacing a hand-crafted MCTS simulation policy by
one based on learned board patterns can increase a player's strength. Following
Go players such as ERICA, we apply a minorization-maximization algorithm [12]
to measure the win correlation of board patterns, and use the resulting pattern
weights in both the leaf selection and simulation phases of MCTS. The result is
MoHex 2.0, a player that is about 250 Elo[1] points stronger than MoHex on
the 11×11 board, and 300 Elo points stronger on the 13×13 board.

---

[1] The Elo gain from win rate r is $400 * -\log((1/r) - 1)$.

H.J. van den Herik et al. (Eds.): CG 2013, LNCS 8427, pp. 60–71, 2014.
DOI: 10.1007/978-3-319-09165-5_6, © Springer International Publishing Switzerland 2014

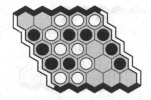

Fig. 1. A Hex game won by White.

In Sect. 2 we review computer Hex. In Sect. 3 we review MoHex. In Sect. 4 we describe MoHex 2.0. In Sect. 5 we give experimental results. In Sect. 6 we discuss future work.

2   Computer Hex

We begin with the rules of Hex. One player has black stones and is assigned two opposite sides, or borders, of the board. The other player has white stones and is assigned the other two sides. Players alternate placing a single stone of their color on an empty cell. The winner is the player who joins their two sides with their stones. Black plays first. To mitigate the first-player advantage, this extra rule is adopted: the first player places a black stone, and the second player then chooses whether to play as white or as black. See Fig. 1. For more on Hex, see [8,9,21].

In the 1950s, as part of his seminal research into games and artificial intelligence, Claude Shannon (with help from E.F. Moore) built electrical circuit machines to play the connection games Birdcage [17] and Hex [29]. Birdcage — also known as Gale or Bridg-It — is similar to Hex, except it is played on the edges of a graph rather than on the nodes [9].

In Shannon's Birdcage circuit, cells have resistors (respectively shorted or removed if occupied by the player or opponent), voltage is applied across the board, and the cell with largest voltage drop is selected as the next move.

This model can also be used for Hex, and cross-board conductance is a good indication of position strength. Following Anshelevich [2], recent alpha-beta search Hex players such as HEXY [1], SIX [19], and WOLVE [4] use evaluations that are based on an augmented circuit that adds information about virtual connections.

A *virtual connection*, or VC, (respectively *virtual semi-connection*, or VSC) is a second-player (first-player) point-to-point connection strategy, where a point is an empty cell, or a *chain* (as in Go, namely a maximal group of stones connected by adjacency), or a board side. Thus a VC between a player's two sides is a winning strategy. Many Hex players find a restricted class of VCs defined by an algebra described by Anshelevich [2]: initialize a set of base VCs, then find larger VCs by applying the closure of two combining rules. The *and-rule* combines VCs in series to form a new VC or VSC. The *or-rule* combines sets of collectively non-interfering VSCs in parallel to form a new VC. See Fig. 2.

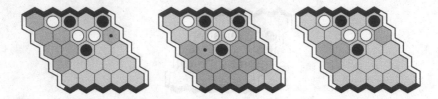

Fig. 2. The left and middle diagrams each show the cells of a side-to-side White VSC. The right diagram shows Black's corresponding mustplay region: a Black move outside this region leaves White with a winning VSC.

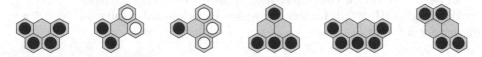

Fig. 3. Some inferior cell patterns. In the first three (from left) patterns, the empty cell is dead and so can be filled for either player. In the last three patterns, the empty cell set is Black-captured and so can be filled for Black.

In addition to strengthening the augmented circuit of which the conductance is the evaluation function, VCs and VSCs are useful in move pruning: if an opponent has a side-to-side VSC, then a player move that fails to interfere with this VSC is losing. With respect to the set of (known) opponent side-to-side VSCs, a player's *mustplay* region is the set of cells that interferes with all of these VSCs. See Fig. 2.

A second form of move pruning is based on inferior moves, especially those deducible from board patterns. A cell is *dead* if it is provably (without changing the win/loss value of the game) never needed by either player. A cell set is *captured* if it can be provably assigned to the capturing player (again, without changing the win/loss value of the game). See Fig. 3. Some Hex players have inferior cell engines, which identify these and other kinds of inferior moves. See [7,23] for more on inferior cell analysis.

3 MoHex

MoHex is a relatively simple first-generation MCTS player [6]. It is built on top of Fuego, an open-source game-independent MCTS platform especially well known for its Go player [15]. MoHex uses UCT (MCTS plus UCB as in [30]) and the RAVE all-moves-as-first UCT heuristic [18]. When a node becomes heavy (i.e., the number of node visits reaches a threshold), it runs its virtual connection and inferior cell engines on that position, yielding information that often prunes inferior moves. In simulations it uses uniformly random moves augmented by only one (up to symmetry) Hex-specific pattern, *savebridge*: if an opponent move attacks a player's bridge (a virtual connection with two empty cells), the player deterministically replies to save the bridge; if the opponent move simultaneously threatens more than one bridge, the player randomly saves one bridge. See Fig. 4.

Fig. 4. The bridge (left). In a MoHEX simulation, when the opponent (White) threatens a bridge, the player (Black) replies to save it (middle, right).

MoHEX is 180 Elo points weaker without RAVE, and a further 100 Elo points weaker without savebridge. For comparison, doubling the number of simulations strengthens MoHEX by 36 Elo points. See [6] for further details.

3.1 MoHEX Weaknesses

Although MoHEX has won three of the past four Olympiad Hex competitions [3,5,6,20], its success is arguably due more to the strength of its virtual connection engine than to its MCTS engine. For example, in recent Olympiad competitions the MCTS players PANORAMEX, MIMHEX, and YOPT each played into winning positions against MoHEX only to play eventually outside of the mustplay region and then lose the game (whenever an opponent misses the mustplay, MoHEX plays out the win without further use of MCTS).

MoHEX is rather weak without its virtual connection and inferior cell engines. This is presumably due to its simulation policy, which is oblivious to global virtual connectivity. In Hex, as in Go, a position can decompose into separate subgames. Consider a position with two subgames in which Black must win both to win the game. If Black's winning probability is .6 for each subgame, then Black's game-winning probability is $\min\{.6, .6\} = .6$, whereas the simulation win rate will be closer to $.6 \times .6 = .36$. Such mismeasurement is a serious flaw, especially against an opponent capable of reasoning about positions with multiple subgames.

Figures 5 and 6 show examples of weak MoHEX performance.

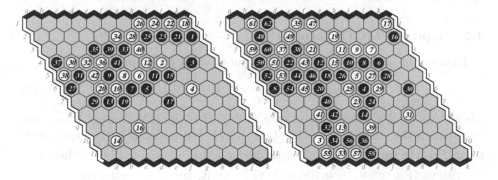

Fig. 5. MoHEX (Black) loses to MIMHEX (left) and WOLVE (right) in 2010.

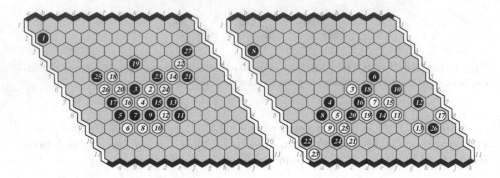

Fig. 6. MoHEX (left) and WOLVE (right) losing to PANORAMEX (White) in the 2011 Olympiad. The latter position is hard for MCTS players: 27.W[j8] connects 13.W[i9] directly to the side but loses; 27.W[i10] connects indirectly to the side and wins.

4 MoHEX 2.0

Given the success of patterns in UCT Go players, we wanted to revise MoHEX by also using patterns. This required substantial revision, so we took the opportunity to add other updates as well. One update is to the virtual connection engine: we now use an algorithm by Jakub Pawlewicz that finds smaller connection sets, but in about half the time. A second update is to tune parameters using the optimization tool CLOP by Remi Coulom [14]. Other changes concern the MCTS engine, as follows.

4.1 Extend on Unstable Search

MCTS is *unstable* if the move with the highest win rate differs from the move with the highest visit count. MoHEX moves are frequently unstable, so we implement a search extension policy similar to that of [25]: extend an unstable search by half of its original search time. We extend a search at most once.

4.2 Improved MCTS Formula

In FUEGO, the score for a move j is computed using the formula

$$Score(j) = (1 - w) \times U + w \times R + E, \tag{1}$$

where U is the UCT mean (wins over visits), R is the RAVE mean (wins over visits), w is the computed weight of the RAVE term, and E is the exploration term. E is the usual UCT exploration formula, i.e., $E = c_b \times \sqrt{\frac{\ln n}{n_j}}$, where n is the parent visit count, n_j is the node visit count, and c_b is a constant.

The RAVE weight w is calculated taking into account the ratio of RAVE visits to UCT visits [18]. For details on FUEGO's RAVE implementation, see [15]. When n_j is small, w is nearly 1, and so $Score(j)$ is almost exactly the

RAVE score. As n_j increases, w decays to 0 and the RAVE score becomes less important.

Notice that the exploration term is separate and outside of the RAVE and UCT terms. Thus large exploration terms (as in the case of a rarely visited child of a popular parent node) can overwhelm the RAVE information, possibly forcing exploration where it is not needed. Indeed, FUEGO and MoHEX both find the best setting of c_b to be 0, so such unneeded exploration never occurs. But a negative side effect of this setting is to rely solely on RAVE for exploration.

As with other recent MCTS players, we find that moving the exploration term inside the UCT term improves performance:

$$Score(j) = (1 - w) \times (U + E) + w \times R. \tag{2}$$

Formula (2) allows exploration (a) to rely on RAVE initially, but (b) to occur for already visited children, which could not happen with formula (1) where $c_b = 0$.

4.3 Patterns

Following Go players such as ERICA, we apply the supervised learning algorithm minorization-maximization (MM) [12] to learn the win correlation of board patterns, and use the resulting pattern weights in both the leaf selection and simulation phases of MCTS.

For learning we used two data sets of about 35 000 games. One set consists of 19 760 13×13 games among strong human players on the Little Golem site, extracted in July 2012. The other set consists of 15 116 games among MoHEX and WOLVE, and also recent Olympiad games.

Most of the computer games in our training data set come from tournaments in which we iterate over all possible opening moves, so the first moves in these games are generally not the strongest available move. Moreover, the human games are played with the swap rule, so the first moves in those games are also not the strongest for that position. Thus, as input for the MM pattern-learning algorithm, for all games we used every move except the first.

The human games often end with a resignation, specifically before either player has an absolute connection between their two sides. By contrast, most of the computer games end with an absolute connection. We obtained the best results when learning on the combined data set rather than on just one of the sets. This suggests that perhaps MoHEX benefits from learning how to play endgames to completion, which is perhaps not surprising given that the simulations have no virtual connection knowledge.

We considered 6-patterns, 12-patterns, and 18-patterns, where a $6t$-pattern consists of the $6t$ cells nearest the pattern center.

In Hex it matters where a player's two sides are. For example, in order to know whether to extend a ladder, a player should know whose side the ladder runs into. Thus, in recording patterns, we allowed only one symmetry, namely rotation by 180 degrees.

Figure 7 shows some patterns learned by MM. In the figures, a is the number of times the pattern appeared, p is the number of times it was played when it was

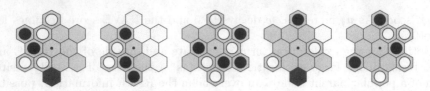

Fig. 7. Learned 12-patterns. Black to move to dot. Shaded cells are board sides. From left: P1 $(\gamma, p, a) = (886, 439, 479)$, P3 $(754, 179, 194)$, P13 $(754, 179, 194)$, P16 $(321, 48, 64)$, P36 $(213, 52, 65)$.

an option, and γ is the feature weight (as defined in [12]) learned by MM, where a larger value is more urgent. Not surprisingly, many of the high-γ patterns are some version of savebridge.

The patterns with high-γ do indeed look urgent. P1, with $(\gamma, p, a) = (886, 439, 479)$, is the pattern with largest γ, played 439 times out of 479: Black splits two White chains and threatens to connect to the Black side. In P3 Black joins two chains and splits the White chain from the White side. In P13 Black joins two chains and splits White chains; this is the highest γ pattern with no side cell. As with most high-γ patterns, this one probably occurs most often near the end of a game. In P16 Black splits White and virtually connects to the Black side via bridges. In P36 Black splits White with a bridge.

Figure 8 shows more patterns. P138 is the 6-pattern with largest γ. Black saves the bridge and splits White from the side.

The default γ-value for moves about which we know nothing is 1.0, so patterns such as P35518, P35474, P35461 — with γ less than .1 — are probably bad moves. P35461 is provably inferior: playing on the other side of the White bridge kills the dotted cell.

A pattern is *global* if it matches anywhere on the board, and *local* if it includes the opponent's most recent move. From our data sets we extracted 65 932 global patterns (33 212 asymmetric: 565 6-patterns, the rest 12-patterns). The maximum γ is 5 820, the minimum is 6.6×10^{-5}. We then ran our inferior cell engine on these patterns; 30 593 of them are *prunable*, namely correspond to a move that is dead, captured, or dominated.

Similarly, we extracted 11 602 local patterns (5 869 asymmetric: 550 6-patterns, the rest 12-patterns), of which 3 659 are prunable. The maximum γ is 11 281, the

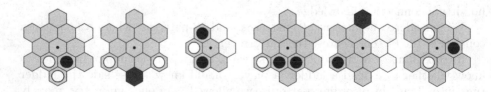

Fig. 8. More learned patterns. P49 (194, 2247, 3259), P135 (100, 86, 182), P138 (98, 94, 191), P35518 (.04, 0, 10190), P35474 (.05, 3, 14270), P35461 (.05, 6, 17351).

minimum is .0262. Prunable patterns corresponding to dead or captured (respectively dominated) moves have γ reset to .00001 (.0001). Patterns that did not occur in the data sets have γ set to 1.0.

4.4 Estimating Prior Knowledge

Following common MCTS practice [18], we use pattern weights as prior knowledge, namely to estimate the relative strengths of a node's unvisited children. When a tree node is visited for the first time, for every empty cell in that position, we identify its unique pattern. If the move is prunable, the move is not considered as a child option for this node. For non-prunable moves, we add the global and local γ-values, and then scale by dividing by the sum of the global and local γ-values over all non-prunable moves.

We use the resulting value as the prior estimated strength value ρ for that move. In addition to the above, an unvisited node has its RAVE value set to .5 and its RAVE count set to 8. Using 12-patterns (respectively 6-patterns, 18-patterns).

4.5 Progressive Bias

Following MCTS practice, e.g., as in Mango [10], we added a progressive bias term to the UCT formula. In our experiments, following Timo Ewalds's Havannah program [16], adding a square root to the denominator of the progressive bias term works best. Here ρ is the prior estimate from Sect. 4.4.

$$Score(j) = (1 - w) \times (U + E) + w \times R + PB. \tag{3}$$

Here $PB = c_{pb} \times \rho/\sqrt{n_j + 1}$, where $c_{pb} = 2.47$ from CLOP.

4.6 Probabilistic Simulations

Having learned the pattern weights, we implemented probabilistic simulations, in which moves are generated stochastically according to a softmax policy [25]. This worked well only after capping the maximum global γ value to a constant, $g = .157$ from CLOP. Thus moves with $\gamma \geq .157$ are equally likely to be played. By contrast, Timo Ewalds (private communication) found no improvement from using probabilistic simulations in the Havannah program CASTRO, possibly because any such improvement is overwhelmed by dynamic factors such as the sudden-death threat from rings.

Notice that resetting the γ values for inferior moves effectively prunes these moves in the simulation without having to call the inferior cell engine.

We are not sure why capping the global γ values is so critical. One possible factor is that the frequency of local moves in the simulations drops if the global values are not capped. In a typical MoHex 2.0 simulation on an empty 13×13 board, about .50 of the moves are local.

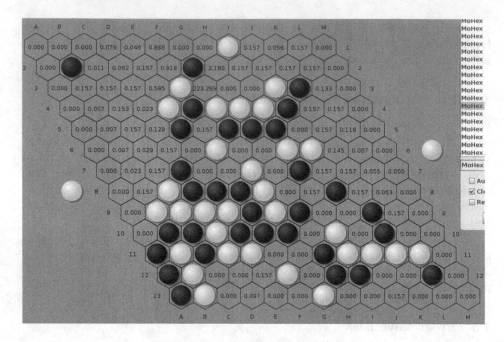

Fig. 9. Part way through a simulation.

Figure 9 is a screenshot from a simulation. The number in a cell is the corresponding move's total γ (local plus global). Notice that many cells have $\gamma = 0$, i.e., less than 0.0005, and so are bad moves or provably inferior, and so unlikely to be selected. The last move played was black g2, which threatens a win at g3. The total γ at g3 is high — much higher than the capped global value of .157 — due to an urgent local pattern — so this move is likely to be selected next. If there was no global cap, this local response would be swamped by global values. It is a coincidence that this response blocks a win threat, because the simulation algorithm knows nothing about global connectivity.

Implementing probabilistic simulations is expensive and requires significant optimization for this feature to not have a negative effect on performance. In MoHex 2.0, this feature increased performance only slightly but was strong enough to overcome a significant (more than a factor of 2) decrease in simulation speed. We chose to keep this feature because we expect it will be useful in future developments.

4.7 Ideas that Did Not Work

In addition to the updates mentioned so far, we considered other changes. Most were not beneficial and so are not included in the final revision. Here are three changes that did not work out.

We were unable to find a stronger hand-crafted simulation policy. Here is a typical attempt: in addition to the 4-cell savebridge pattern, include 4-cell patterns for breakbridge (break the connection when the opponent's last move leaves a partly broken bridge) and ladder continuation. For this set of three patterns, a promising win rate of .6 at 10 K simulations per move dropped to .5 at 100 K per move.

Degrading RAVE weight by distance to the last move, as in FUEGO [24], did not work. Adding criticality (a measure of a cell's importance to the winner [13], perhaps less important in Hex than in Go) did not work.

5   Experimental Results

We ran 11×11 and 13×13 tournaments between MoHEX 2.0 and the 2011 Olympiad version of MoHEX. Each tournament iterated multiple times over all possible opening moves. Each player had 4 cores, 1.5 Gb memory, and time per game up to 5 min. Each thread ran MCTS, so neither player used the single-thread solver that MoHEX has run in recent Olympiads. Over more than 1000 games, MoHEX 2.0 had respective win rates of .81 and .85, meaning again of about 252 and 310 Elo points, respectively. See Fig. 10.

board size	allotted time per player		
	1 min	3 min	5 min
11×11		.811 ± .010	
13×13	.853 ± .006	.852 ± .006	.856 ± .010

Fig. 10. MoHEX 2.0 v MoHEX win rates. ± is standard error, 68 % confidence.

We also ran a 3000-game 3 min/game 13×13 tournament comparing these programs against the current version of WOLVE. The MoHEX and MoHEX 2.0 win rates are .587±.008 and .854±.006 respectively. This is a 245 Elo points gain, showing that the improvements to MoHEX are not just an artefact of self-play (Fig. 11).

6   Future Work

While MoHEX 2.0 is stronger than MoHEX, it is still far from perfect. For example, it does not find the winning move from the PANORAMEX-WOLVE game shown in Fig. 6. There are several things to try next. We mention four of them. (1) Larger patterns could be used, both in the tree and simulations. (2) Global connectivity information could be added to the simulations, which currently know nothing about winning or losing even when only one move away. (3) Simulation balancing could be used to learn simulation weights [25], which are currently those learned by MM and also used in the tree. (4) Adaptive simulation policies could be used.

program/feature	Elo gain
MoHex 1.0	—
new VC algorithm/MCTS formula	*
extend on unstable search	35
patterns in tree	175
probabilistic simulations	*
CLOP	46
total: MoHex 2.0	310

Fig. 11. Feature contributions to MoHex 2.0 on 13×13 board. * entries are not easily measurable in current framework.

References

1. Anshelevich, V.V.: Hexy wins Hex tournament. ICGA J. **23**(3), 181–184 (2000)
2. Anshelevich, V.V.: A hierarchical approach to computer Hex. Artif. Intell. **134**(1–2), 101–120 (2002)
3. Arneson, B., Hayward, R.B., Henderson, P.: Mohex wins Hex tournament. ICGA J. **32**(2), 114–116 (2009)
4. Arneson, B., Hayward, R.B., Henderson, P.: Wolve 2008 wins Hex tournament. ICGA J. **32**(1), 49–53 (2009)
5. Arneson, B., Hayward, R.B., Henderson, P.: Mohex wins Hex tournament. ICGA J. **33**(3), 181–186 (2010)
6. Arneson, B., Hayward, R.B., Henderson, P.: Monte-Carlo tree search in Hex. IEEE Trans. Comput. Intell. AI Games **2**(4), 251–258 (2010)
7. Björnsson, Y., Hayward, R.B., Johanson, M., van Rijswijck, J.: Dead cell analysis in Hex and the Shannon game. In: Bondy, A., Fonlupt, J., Fouquet, J.-L., Fournier, J.-C., Ramirez Alfonsin, J.L. (eds.) Graph Theory in Paris: Proceedings of a Conference in Memory of Claude Berge, pp. 45–60. Birkhäuser (2007)
8. Browne, C.: Hex Strategy: Making the Right Connections. A.K. Peters, Natick (2000)
9. Browne, C.: Connection Games: Variations on a Theme. A.K. Peters, Wellesley (2005)
10. Chaslot, G.M.J.-B., Winands, M.H.M., van den Herik, H.J., Uiterwijk, J.W.H.M., Bouzy, B.: Progressive strategies for monte-carlo tree search. New Math. Nat. Comput. (NMNC) **04**(03), 343–357 (2008)
11. Coulom, R.: Efficient selectivity and backup operators in Monte-Carlo tree search. In: van den Herik, H.J., Ciancarini, P., Donkers, H.H.L.M.J. (eds.) CG 2006. LNCS, vol. 4630, pp. 72–83. Springer, Heidelberg (2007)
12. Coulom, R.: Computing "elo ratings" of move patterns in the game of go. ICGA J. 30(4), 198–208 (2007). http://remi.coulom.free.fr/Amsterdam2007/
13. Coulom, R.: Criticality: a monte-carlo heuristic for go programs (2009). http://remi.coulom.free.fr/Criticality/
14. Coulom, R.: CLOP: confident local optimization for noisy black-box parameter tuning. In: 13th Advances in Computer Games, pp. 146–157 (2011). http://remi.coulom.free.fr/CLOP/
15. Enzenberger, M., Müller, M., Arneson, B., Segal, R., Xie, F., Huang, A.: Fuego (2007–2012). http://fuego.sourceforge.net/

16. Ewalds, T.: Playing and solving havannah. Master's thesis, University of Alberta (2012)
17. Gardner, M.: Recreational topology. In: The 2nd Scientific American Book of Mathematical Puzzles and Diversions, Chap. 7, pp. 78–88. Simon and Schuster, New York (1961)
18. Gelly, S., Silver, D.: Combining online and offline knowledge in UCT. In: Ghahramani, Z. (ed.) ICML. ACM International Conference Proceeding Series, vol. 227, pp. 273–280. ACM (2007)
19. Hayward, R.B.: Six wins Hex tournament. ICGA J. **29**(3), 163–165 (2006)
20. Hayward, R.B.: Mohex wins Hex tournament. ICGA J. **36**(3), 180–183 (2013)
21. Hayward, R.B., van Rijswijck, J.: Hex and combinatorics. Discrete Math. **306** (19–20), 2515–2528 (2006)
22. Hein, P.: Vil de laere Polygon? Politiken, 26 December 1942
23. Henderson, P.: Playing and Solving the Game of Hex. Ph.D. thesis, University of Alberta, Edmonton, Alberta, Canada (2010)
24. Huang, S.-C.: New Heuristics for Monte Carlo Tree Search Applied to the Game of Go. Ph.D. thesis, Nat. Taiwan Normal Univ., Taipei (2011)
25. Huang, S.-C., Coulom, R., Lin, S.-S.: Time management for Monte-Carlo tree search applied to the game of go. In: 2010 International Conference on Technologies and Applications of Artificial Intelligence (TAAI), pp. 462–466, November 2010
26. Kuhn, H.W., Nasar, S. (eds.): The Essential John Nash. Princeton University Press, Princeton (2002)
27. Nasar, S.: A Beautiful Mind: A Biography of John Forbes Nash. Jr, Simon and Schuster (1998)
28. Nash, J.: Some games and machines for playing them. Technical report D-1164, RAND, February 1952
29. Shannon, C.E.: Computers and automata. Proc. Inst. Radio Eng. **41**, 1234–1241 (1953)
30. Wang, Y., Gelly, S.: Modifications of uct and sequence-like simulations for monte-carlo go. In: CIG, pp. 175–182. IEEE (2007)

Analyzing Simulations in Monte-Carlo Tree Search for the Game of Go

Sumudu Fernando and Martin Müller[✉]

University of Alberta, Edmonton, Canada
{sumudu,mmueller}@ualberta.ca

Abstract. In Monte-Carlo Tree Search, simulations play a crucial role since they replace the evaluation function used in classical game-tree search and guide the development of the game tree. Despite their importance, not too much is known about the details of how they work. This paper starts a more in-depth study of simulations, using the game of Go, and in particular the program FUEGO, as an example. Playout policies are investigated in terms of the number of blunders they make, and in terms of how many points they lose over the course of a simulation. The result is a deeper understanding of the different components of the FUEGO playout policy, as well as an analysis of the shortcomings of current methods for evaluating playouts.

1 Introduction

While Monte-Carlo Tree Search (MCTS) methods are extremely successful and popular, their fundamentals have not been understood and tested well. In MCTS, statistics over the win rate of randomized simulations or "rollouts" are used as a state evaluation. One unstated but "obvious" assumption is that the quality of the simulation policy is closely related to the performance of the resulting search.

How to measure such quality? So far, three main studies have addressed this question. Gelly and Silver [7] note that a playout policy which is stronger when used as a stand-alone player does not necessarily work better when used as a simulation policy in an MCTS player. Silver and Tesauro's experiments with small-board Go [12] indicate that it is more important that a simulation policy is unbiased (or balanced) rather than strong. Huang, Coulom, and Lin [8] develop this idea further and create a practical training algorithm for full-scale Go.

This paper aims to re-open the study of simulation policies for MCTS, using the game of Go, and in particular the program FUEGO [6], as a test case. Section 2 introduces notation, hypotheses, and research questions. Section 3 describes the methods used in this investigation, Sect. 4 presents the data obtained, and Sect. 5 outlines the conclusions drawn. Section 6 suggests future work to extend these results.

H.J. van den Herik et al. (Eds.): CG 2013, LNCS 8427, pp. 72–83, 2014.
DOI: 10.1007/978-3-319-09165-5_7, © Springer International Publishing Switzerland 2014

2   Some Hypotheses About Playout Policies

The following notation will be used throughout this paper: $\mathcal{P}$ denotes a playout policy. MCTS($\mathcal{P}$) is an MCTS player using policy $\mathcal{P}$ in its playouts. In the experiments, specific policies and MCTS players will be introduced.

A *blunder* can be defined as a game-result-reversing mistake. A simulation starting in position p gives the game-theoretically correct result if it plays an even number of blunders. For example, if neither player blunders, then the result of the game at the end of the simulation is the same as the game-theoretic result obtained with best play by both sides from p. If there is exactly one blunder in the simulation, then at some stage the player who was winning at p makes a crucial mistake, and is losing from that point until the end of the simulation.

The first two of the following hypotheses are investigated in this work.

1. *The strength of a playout for an MCTS engine correlates strongly with whether it preserves the win/loss of the starting position.* The correlation between evaluation before and after the playout is the most important factor in quality of playout. In Sect. 5.2, we report an experiment that approximately counts the number of blunders in many policies $\mathcal{P}_i$, and checks if it correlates with the playing strength of MCTS($\mathcal{P}_i$). Interestingly, the result is negative. We give a partial explanation.
2. *The size of errors made during simulation matters.* Our approach estimates the score for each position by a series of searches with varying komi. Then compare the estimated score before and after each move. Such an experiment is performed in Sect. 5.3, with promising results.
3. *Considering simulation balance, having no systematic bias is much more important than low error.* This hypothesis is addressed by the work on simulation balancing [8, 12]. The quality of a playout can be described by bias and variance. Bias is the most important factor; after that, less variance is better.

3   Methodology

All experiments were conducted using FUEGO SVN revision 1585, with some modifications outlined below, on a 9×9 board. When using MCTS, FUEGO was configured to ignore the clock and to perform up to 5000 simulations per search. The opening book was switched off.

For a download from the FUEGO wiki are available: (1) detailed FUEGO configuration files, (2) the source code patches required to reproduce these experiments, and (3) the raw data produced by this investigation at http://sourceforge. net/apps/trac/fuego/wiki/SimulationPolicies.

3.1   Playout Policies in FUEGO

The playout policy used by Fuego to simulate games from a leaf node of the MCTS tree is structured into several subpolicies which are applied in a top-down manner: the move to play is chosen from the set of moves returned by

Table 1. Policy variations used in this investigation.

$\mathcal{F}$	Default policy	$\mathcal{R}$	Pure random policy
Subtractive policies		*Additive* policies	
no_ac	$\mathcal{F}$ − AtariCapture	only_ac	$\mathcal{R}$ + AtariCapture
no_ad	$\mathcal{F}$ − AtariDefense	only_ad	$\mathcal{R}$ + AtariDefense
no_bp	$\mathcal{F}$ − BiasedPatternMove	only_bp	$\mathcal{R}$ + BiasedPatternMove
no_patt	$\mathcal{F}$ − PatternMove	only_patt	$\mathcal{R}$ + PatternMove
no_ll	$\mathcal{F}$ − LowLib	only_ll	$\mathcal{R}$ + LowLib
no_cap	$\mathcal{F}$ − Capture	only_cap	$\mathcal{R}$ + Capture
no_fe	$\mathcal{F}$ − FalseEyeToCapture		

the first subpolicy that proposes a non-empty set of moves. This choice is made uniformly at random, except in the case of the BiasedPatternMove subpolicy which weights moves according to machine-learned probabilities [5]. The final selected move may also be adjusted by one or more move correctors which are designed to replace certain "obviously" suboptimal moves. For specific details of FUEGO's playout policy and how it works together with other MCTS components in FUEGO, please refer to [5,6].

In this investigation, variations of the default FUEGO policy were obtained by activating subsets of the available subpolicies, as summarized in Table 1. In addition to the default policy (denoted $\mathcal{F}$) and the purely random (except for filling eyes) policy (denoted $\mathcal{R}$), the policy variations are divided into the *subtractive* policies (obtained by disabling one subpolicy from $\mathcal{F}$) and the *additive* policies (obtained by adding one subpolicy to $\mathcal{R}$). The subtractive policies also include the "no_fe" policy obtained by disabling the FalseEyeToCapture move corrector in $\mathcal{F}$. Note also that the "no_patt" policy excludes both unbiased and biased pattern moves.

The *gogui-twogtp* utility from the *GoGui* suite [4] was used to establish the relative strengths of MCTS($\mathcal{P}$) for all $\mathcal{P}$ in Table 1, via self-play. Specifically, each subtractive MCTS($\mathcal{P}$) was played against MCTS($\mathcal{F}$) 2000 times, and each additive MCTS($\mathcal{P}$) was played against MCTS($\mathcal{R}$) 500 times.

3.2 Blunders and Balance

Following Silver and Tesauro [12], the playing strength of MCTS($\mathcal{P}$) can be related to the *imbalance* of the playout policy $\mathcal{P}$. In plain terms, a playout policy that tends to preserve the winner of a game can be expected to produce good performance in MCTS, even if the moves played during the policy's simulation are weak.

To investigate this idea, each subtractive policy $\mathcal{P}$ (the additive policies proved too weak) was played against itself over 100 games (again using *gogui-twogtp*). Every position in the game was evaluated using MCTS($\mathcal{F}$), yielding an estimate (UCT value) of Black's probability of winning the game assuming

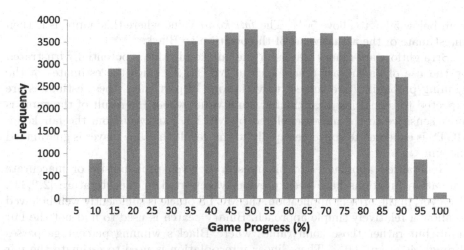

Fig. 1. Distribution of sampled positions.

strong play from that point. Ignoring the first 10 moves, any policy move producing a swing in UCT value from over 0.75 to under 0.25 (or vice-versa for White) was flagged as a *blunder*. This approach to empirical approximation of a blunder is similar to the one taken in [12].

3.3 Sampling Positions from Real Games

In order to obtain positions representative of actual games, random samples were taken from real 9×9 Go games played on CGOS [3]. Starting with the 57,195 games from June 2011.

- All games shorter than 20 moves were removed, leaving 56,968 games.
- The remaining games were labeled in increasing order of CGOS sequence number: for example, game 1,616,263 is labeled 1, game 1,616,264 is labeled 2, etc.
- From the game labeled i having n_i moves in total, the position after $10 + (i \bmod (n_i - 19))$ moves was saved as a sample. This excludes positions closer than 10 moves to either end of the game.

As shown in Fig. 1, this sampling procedure produces a reasonable spread of early-game, mid-game, and late-game positions, while also being easy to reproduce using the publicly available CGOS records.

3.4 Quantifying Policy Errors

Given a Go position including the player whose turn it is, a strong MCTS player such as MCTS($\mathcal{F}$) can be used as an oracle to determine Black's winning probability. Assuming that the oracle supports variation of the komi value, a binary search can be used to find the point at which Black's chances of winning change from below 50 % to above 50 %. The *fair komi* value where this happens is then an estimate of the actual value of the position to Black.

Straightforward binary search for the fair komi value is potentially frustrated by the use of an inexact oracle such as MCTS($\mathcal{F}$) since the estimates of the winning probability are subject to variation. In particular, these estimates are expected to be noisiest near the fair komi value where the result of the game is most sensitive to the moves explored by MCTS. Far away from the fair komi, MCTS is expected to converge rapidly to the result that one player is guaranteed the win.

Generalized approaches to binary search given probabilistic or inaccurate evaluation functions have been previously reported in the literature [2,9,11]. However, in the present application, the MCTS oracle is sufficiently well-behaved to allow a relatively simple approach. Binary search is used to find not the fair komi, but rather those komi values where Black's winning percentage passes through 40 % and 60 %. Then, linear interpolation is used to estimate the fair komi corresponding to a winning percentage of 50 %.

For each position in the CGOS sample set, an estimate of its value was obtained using the procedure outlined above. Then, every legal move from the position was played and the value estimate for the resulting position computed: the change in position value is an estimate for the value of the move itself. These move ratings were saved as an "annotated" position file to allow for quickly rating any play from the position. Using MCTS($\mathcal{F}$) with 5000 simulations as an oracle, a typical middle game position takes about 5 min to annotate using this method. This functionality was temporarily added to the FUEGO code itself in order to simplify the processing of the sample set.

Using the annotated positions from the test set, the play of each policy variation $\mathcal{P}$ was evaluated quantitatively. Since $\mathcal{P}$ is non-deterministic, new functionality was added to FUEGO to return a list of all possible policy moves, along with their relative weighting, rather than a single randomly chosen move. This permitted the accurate calculation of the expected value of the policy's play from an annotated position.

In this setting, we redefine our approximation of a blunder as a move that changes a winning position into a loss, and that loses more than some set number of points. A threshold of 5 points was chosen based on a qualitative estimate of the precision of the move values estimated by the fair komi method.

4 Results

The results are partitioned into three subsections: Playout policies (Sect. 4.1), Blunders and balance (Sect. 4.2), and Quantifying policy errors (Sect. 4.3).

4.1 Playout Policies in FUEGO

The statistics from policy self-play in Table 2 illustrate the relative effectiveness of the various subpolicies. In particular, the PatternMove and AtariDefense subpolicies are clearly rather effective based on the subtractive results. Also, the

Table 2. Relative strengths of MCTS($\mathcal{P}$), estimated from 2000 (left column) and 500 (right column) games respectively. Standard error as reported by *gogui-twogtp*.

$\mathcal{P}$	Win % vs. MCTS($\mathcal{F}$)	$\mathcal{P}$	Win % vs. MCTS($\mathcal{R}$)
$\mathcal{F}$	49.5 ± 1.1	$\mathcal{R}$	48.3 ± 2.2
no_ac	51.6 ± 1.1	only_ac	57.6 ± 2.2
no_ad	18.2 ± 0.9	only_ad	70.3 ± 2.0
no_bp	46.4 ± 1.1	only_bp	76.0 ± 1.9
no_patt	25.9 ± 1.0	only_patt	66.8 ± 2.1
no_ll	47.5 ± 1.1	only_ll	59.1 ± 2.2
no_cap	50.4 ± 1.1	only_cap	67.1 ± 2.1
no_fe	49.9 ± 1.1		

Table 3. Blunder statistics for subtractive policy self-play, 100 games.

$\mathcal{P}$	Blunder-free games	Even-blunder games	% of blunder moves
$\mathcal{F}$	53	72	3.00
no_ac	62	82	2.78
no_ad	69	85	2.14
no_bp	55	77	2.95
no_patt	38	65	4.15
no_ll	52	79	4.25
no_cap	56	71	2.67
no_fe	69	84	1.70

BiasedPatternMove subpolicy stands out among the additive results. Of note is the fact that disabling the AtariCapture subpolicy resulted in a slight but statistically significant advantage as compared to the default configuration.

4.2 Blunders and Balance

Blunder statistics from policy self-play are summarized in Table 3. In addition to the raw percentage of blunder moves, both blunder-free games and games with an even number of blunders were counted. If a policy tends to play an even (including zero) number of blunders, then though these moves are objectively weak the policy can still be effectively used in MCTS, since the blunders will cancel out and the policy will still preserve the winner of a game.

4.3 Quantifying Policy Errors

Figure 2 shows that a linear approximation to the neighbourhood of the fair komi is reasonably accurate, even for early-game positions. Figure 3 plots the fair komi estimate vs. move number for the CGOS samples, confirming that the sample set includes both one-sided and contested positions over a variety of game stages.

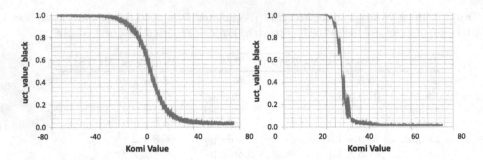

Fig. 2. MCTS oracle values vs. komi for early-game position (left: 1616349 after 10 moves) and mid-game position (right: 1616808 after 45 moves).

Table 4. Expected policy performance (per move, averaged over all CGOS samples).

$\mathcal{P}$	Blunder chance (%)	Expected point loss
$\mathcal{F}$	4.14	4.52
no_ac	4.19	4.55
no_ad	4.67	4.81
no_bp	4.51	4.73
no_patt	6.55	5.57
no_ll	4.10	4.47
no_cap	4.20	4.62
no_fe	4.14	4.53
$\mathcal{R}$	9.95	7.48
only_ac	9.42	7.12
only_ad	8.22	6.46
only_bp	4.77	4.97
only_patt	5.82	5.55
only_ll	8.40	6.75
only_cap	8.63	6.61

Table 4 summarizes the data collected by evaluating policy moves on the annotated CGOS samples. Recall that in this context, a blunder was defined as a move from a winning position to a losing position that also loses at least 5 points (according to the fair komi estimate).

5   Discussion and Suggestions

In this section we discuss three topics and provide seven suggestions for a closer investigation, viz. Playout policies in FUEGO (Sect. 5.1, one suggestion), Blunders and balance (Sect. 5.2, two suggestions), and Quantifying policy errors (Sect. 5.3, four suggestions).

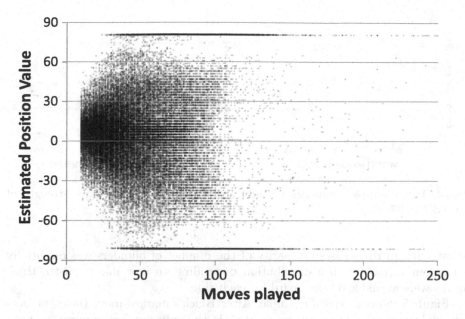

Fig. 3. Fair komi estimate vs. move number in sampled positions. Darker regions represent more samples.

5.1 Playout Policies in FUEGO

The results summarized in Table 2 were for the most part as expected, although the small improvement obtained when disabling the AtariCapture subpolicy was interesting. A possible explanation for this is that the AtariCapture subpolicy sometimes eagerly suggests poor moves, while most or all of the good moves it finds would be covered by subpolicies later in the chain (e.g., [Biased]Pattern Move) — in this case disabling the AtariCapture subpolicy (suggestion 1) could improve the accuracy of playouts. However, doing the disabling increases the average playout time, since AtariCapture is (i) a simpler subpolicy than Biased-PatternMove, and (ii) is not effective in timed games.

5.2 Blunders and Balance

As shown in Fig. 4, there was no appreciable correlation between the blunder statistics obtained from self-play of policy $\mathcal{P}$ and the relative strength of MCTS($\mathcal{P}$) for subtractive $\mathcal{P}$. It appears that another quite nuanced approach is needed to test the idea that more balanced policies are more effective for MCTS. One obvious deficiency with the present method is that the positions arising in pure self-play policy differ greatly from the positions at the bottom of a typical MCTS tree. The deficiency could be addressed by initiating the policy by self-play from "real-world" positions instead of an empty board (suggestion 2). A more critical issue, however, was the observed occurrence of long series of blunders in which the policy repeatedly missed a crucial play for both sides (which still *is* balanced

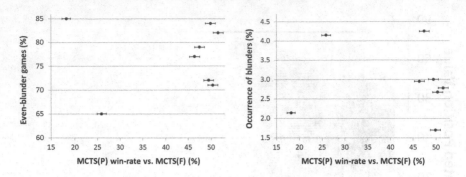

Fig. 4. Policy self-play statistics for subtractive policies $\mathcal{P}$ vs. relative strength of MCTS($\mathcal{P}$) (100 playouts)

behaviour). In these cases the parity of the number of blunders was essentially a random variable with a distribution depending on how many "distracting" moves were available on the board for each side.

Figure 5 shows a typical example. After Black's marked move (near the bottom right corner), the nakade move at A is the only non-losing move for both sides. Without specialized nakade knowledge in the playouts to generate that play (suggestion 3: add nakade knowledge), it is up to chance whether Black will kill or White will make two eyes. In such situations it is difficult to ascribe a significant meaning to the statistic "chance of committing an even number of blunders". The possibility of many smaller, non-blunder moves adding up to a game-reversing result also serves to frustrate the analysis focused on blunders. Predicting the strength of an MCTS player based on its policy's self-play behaviour is significantly more difficult than simply examining result-reversing moves.

5.3   Quantifying Policy Errors

Good estimates of the errors in move values estimated by the fair komi method are still needed. Simply repeating the annotation of a given position reliably produced variations in move values of at most a few points, suggesting that MCTS($\mathcal{F}$) using 5000 simulations produces reasonably precise (if not accurate) values. Time permitting, increasing the number of simulations could confirm and/or improve the current estimates. Furthermore, using other MCTS engines as oracles would provide material for an interesting comparative study (suggestion 4); positions where estimates differ greatly could prove to be interesting test cases.

Figure 6 depicts statistics collected by analysing policy behaviour on the CGOS sample positions. First, a strong correlation between the expected point loss and the policy blunder rate was evident. This correlation is intuitively appealing, as it affirms that the occurrence of blunder moves (which were only a small portion of all the moves examined) can be linked to the strength of a

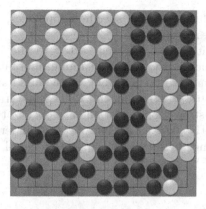

Fig. 5. A nakade position leading to a series of "blunders".

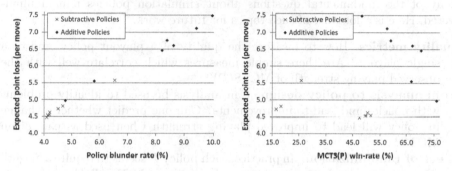

Fig. 6. Expected point loss of policy $\mathcal{P}$ vs. policy blunder rate (left) and vs. relative strength of MCTS($\mathcal{P}$) (right).

policy even in non-blunder situations. Second, in comparing the expected point loss of policy $\mathcal{P}$ to the relative strength of MCTS($\mathcal{P}$), a suggestive negative correlation emerged, particularly for the additive policies (note that relative strength was measured against MCTS($\mathcal{F}$) for subtractive $\mathcal{P}$ and against MCTS($\mathcal{R}$) for additive $\mathcal{P}$: refer to Subsects. 3.1 and 4.1). This data supports the idea that the expected point loss of a policy $\mathcal{P}$ can be used as a predictor (though imperfect) of the strength of MCTS($\mathcal{P}$) (suggestion 5). Of course, it is also evident that expected point loss is far from being the only factor.

The annotation and subsequent evaluation of policy moves on the CGOS sample positions has produced a wealth of quantitative data, but it is not yet clear how to use this data to improve policies for use in MCTS players such as FUEGO. Among other ideas meriting further investigation, using the detailed statistics to optimize the gross structure of the playout procedure is particularly attractive (suggestion 6). For example, at different stages of the game the order of evaluating subpolicies may be varied (or the subpolicies may be given different weights), guided by the expected point losses. More focused uses could include identifying pathological behaviour in specific subpolicies, and developing filters to improve them (suggestion 7).

The annotated position files are well suited to the iterative development and/or testing of policies as the time-consuming annotation only needs to be done once (except to improve the accuracy of the annotations). They can also serve as fine-grained regression tests for full MCTS players, though the problem of identifying "ideal" test positions is not an easy one. Out of the 56968 sampled positions, MCTS($\mathcal{F}$) with 5,000 simulations plays a blunder in 242 of them, which indicates that examining these positions in closer detail may help to identify weaknesses in MCTS($\mathcal{F}$).

6   Future Work

Many of the fundamental questions about simulation policies remain unanswered. Here is a long list of such topics for future work.

Quality metrics. How to measure the quality of a playout policy, and of a single playout? Are there simple measures which correlate well with the observed playing strength of MCTS($\mathcal{P}$)?

From analysis to policy design. Can analysis be used to identify problems with policies and lead to improvements? Can one predict whether a change in policy will lead to improved playing strength when used as part of an MCTS?

Effect of randomization. In practice, each policy $\mathcal{P}$ seems to require a "right" amount of randomization, and the performance of MCTS($\mathcal{P}$) decreases outside the optimum. Can we develop a theoretical understanding of why and how randomization works, and how to choose the correct amount?

Local vs. global error. Is accurate full-board win/loss estimation more important than detailed point evaluation? Can local errors in $\mathcal{P}$ cancel each other without affecting the playing strength of MCTS($\mathcal{P}$)?

Pairs of policy moves. Silver and Tesauro [12] addressed the idea of two-step imbalance, which could be explored with our approach. The idea is to play two policy moves, then use binary search to estimate the resulting two-step imbalance.

Evaluation using perfect knowledge. Policies can be evaluated 100 % correctly if a domain-specific solver is available. Examples are Hex, where very strong general endgame solvers are available [1], and in a more restricted sense Go, where endgame puzzles with provably safe stones, which divide the board into sufficiently small regions, can be solved exactly using decomposition search [10].

Policy vs. short search. Instead of just evaluating the move proposed by the policy $\mathcal{P}$ itself, it would be interesting to evaluate the move played by MCTS($\mathcal{P}$) with a low number of simulations, and compare with $\mathcal{P}$'s move.

Game phases. Analyze the behavior of policies in different game phases, and design more fine-grained policies for each phase.

Other programs and other games. Try similar experiments with other programs such as PACHI in Go, or even other games such as Hex.

Acknowledgements. Support for this research was provided by NSERC, and the Departments of Electrical and Computer Engineering and of Computing Science at University of Alberta.

References

1. Arneson, B., Hayward, R., Henderson, P.: Solving hex: beyond humans. In: van den Herik, H.J., et al. [13], pp. 1–10 (2011)
2. Ben-Or, M., Hassidim, A.: The Bayesian learner is optimal for noisy binary search (and pretty good for quantum as well). In: FOCS, pp. 221–230. IEEE, Washington, DC (2008)
3. Dailey, D.: Computer Go Server (2007–2013). http://cgos.boardspace.net. Accessed 10 April 2013
4. Enzenberger, M.: GoGui (2007–2013). http://gogui.sourceforge.net. Accessed 10 April 2013
5. Enzenberger, M., Müller, M.: Fuego homepage (2008–2013). http://fuego.sf.net. Accessed 10 April 2013
6. Enzenberger, M., Müller, M., Arneson, B., Segal, R.: Fuego - an open-source framework for board games and Go engine based on Monte Carlo tree search. IEEE Trans. Comput. Intell. AI Games **2**(4), 259–270 (2010)
7. Gelly, S., Silver, D.: Combining online and offline knowledge in UCT. In: ICML '07: Proceedings of the 24th International Conference on Machine Learning, pp. 273–280. ACM (2007)
8. Huang, S., Coulom, R., Lin, S.: Monte-Carlo simulation balancing in practice. In: van den Herik, H.J., et al. [13], pp. 81–92 (2011)
9. Karp, R.M., Kleinberg, R.: Noisy binary search and its applications. In: SODA, pp. 881–890. SIAM, Philadelphia, PA, USA (2007)
10. Müller, M.: Decomposition search: a combinatorial games approach to game tree search, with applications to solving Go endgames. In: IJCAI, pp. 578–583, Stockholm, Sweden (1999)
11. Nowak, R.: Generalized binary search. In: Proceedings of the 46th Allerton Conference on Communications, Control, and Computing, pp. 568–574 (2008)
12. Silver, D., Tesauro, G.: Monte-Carlo simulation balancing. In: Danyluk, A., Bottou, L., Littman, M. (eds.) ICML, vol. 382, pp. 945–952. ACM (2009)
13. van den Herik, H.J., Iida, H., Plaat, A. (eds.): CG 2010. LNCS, vol. 6515. Springer, Heidelberg (2011)

Anomalies of Pure Monte-Carlo Search in Monte-Carlo Perfect Games

Ingo Althöfer[✉] and Wesley Michael Turner

Faculty of Mathematics and Computer Science, Friedrich-Schiller-University,
Ernst-Abbe-Platz 2, 07743 Jena, Germany
ingo.althoefer@uni-jena.de

Abstract. A game is called "Monte-Carlo perfect" when in each position pure Monte-Carlo search converges to perfect play as the number of simulations tends toward infinity. We exhibit three families of Monte-Carlo perfect single-player and two-player games where this convergence is not monotonic. We for example give a class of MC-perfect games in which MC(1) performs arbitrarily well against MC(1,000).

1 Introduction

When discussing pure Monte-Carlo search, an agent with parameter k generates and records the scores for k random games beginning with each of the position's feasible moves. The move with the best average score is then played. Ties between any best-scorers are broken by a fair random decision. We call a game "Monte-Carlo perfect" [1] when in each of a game's possible positions this simple procedure converges to perfect play as k goes to infinity.

Most games are not Monte-Carlo perfect. Explicit examples may be found in [2,3]. In such cases, it may be appropriate to embed Monte-Carlo in a tree-search framework. See, for instance, MCTS [4] and UCT [5].

But strange things can happen in Monte-Carlo perfect games, too. Convergence to perfect play does not mean that playing strength increases monotonically with the k parameter. In this paper we exhibit several classes of Monte-Carlo perfect single-player and two-player games where convergence to optimal performance is not monotone. For some of these examples we give exact evaluations of concrete games. In others, an abstract analysis of whole game classes.

The main finding: For each pair (i, k) of integers with $i < k$ and each $\epsilon \in (0, 1]$ there exists a Monte-Carlo perfect game $G(i, k; \epsilon)$ where the winning probability of $MC(i)$ against $MC(k)$ is $1 - \epsilon$ or better.

The paper is organized as follows. In Sect. 2 we present some short Monte-Carlo perfect games called *Double Step Races with Black Holes* in which we first found an occurrence of Monte-Carlo agents with smaller parameters scoring more than fifty percent in contests with agents with larger parameters. Sections 3 and 4 give existence theorems for families of two-player and for single-player games respectively and the basics for their construction. Section 5 contains some concrete examples (short games with relatively few positions) where "smaller" MC

H.J. van den Herik et al. (Eds.): CG 2013, LNCS 8427, pp. 84–99, 2014.
DOI: 10.1007/978-3-319-09165-5_8, © Springer International Publishing Switzerland 2014

agents achieve surprising scores against their bigger brothers. Since in some sense the constructions in Sect. 5 are simpler than those in the previous sections, for some readers it may make sense to look at the parts in reversed order. Section 6 explores the presence of anomalies in very small game trees, and finally Sect. 7 concludes with a discussion and six open questions.

Readers should keep two things in mind.

- The games analyzed in this paper are not meant for playing in the real world. Their only purpose is to demonstrate the counter-intuitive non-monotonic playing strength of pure Monte-Carlo in MC perfect games.
- For those with a mathematical background, it is not difficult to verify the properties claimed for our games. But it was hard work to find the original games and simplify them to be as clearly structured as they are now. Technique in perfection often becomes unobtrusive!

2 Double Step Races with Black Holes

Below we discuss two versions of Double Step Races (DSR) with Black Holes (Family 1). In Subsect. 2.1 we investigate Length 3 Double Step Races with 16 Black Holes (Family 1, class 1), and in Subsect. 2.2 we briefly review Length 3 Double Step Races with 47 Black Holes (Family 1, class 2). In Subsect. 2.2 we also report on further increasing the Black Holes (Family 1, class 3).

2.1 Length 3 Double Step Races with 16 Black Holes

Double Step Races are a trivial family of games for humans to play, but useful test-beds for analyzing properties of pure Monte-Carlo search (see [6–8]). In the simplest version of this game, each player has one lane of a given length and a single piece that "runs" down this lane. Moves are made in one direction only (from left to right), and the feasible moves are to move the piece either one or two steps down the lane (a single step or double step, respectively). The first player to reach the rightmost cell of his[1] lane is the winner. An intelligent player will of course make double steps all the time, except sometimes on the final move when there is only a single step left.

"DSR-3 with 16 Black Holes" is a class within "Double Step Race 3" (class 1). Each player has a lane of length 3 (4 spaces if the starting space is included) and one piece in this lane. In addition, each player has a second piece which has 16 feasible moves, each of which brings this "black-hole" piece into a black hole from which it will never return. As in the base version of the game, the first player to reach the target cell in his lane of length 3 wins the game.

Again, for an intelligent player it is clear that he will reach the end in the fewest number of moves (namely two) by only ever moving the piece in his lane. The second piece and its black holes are included only to make the situation complicated for Monte-Carlo agents.

---

[1] For Brevity, we use 'he' and 'his' whenever 'he or she' and 'his or her' are meant.

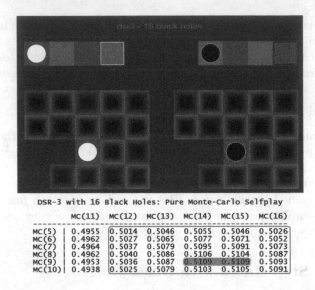

DSR-3 with 16 Black Holes: Pure Monte-Carlo Selfplay

	MC(11)	MC(12)	MC(13)	MC(14)	MC(15)	MC(16)
MC(5)	0.4955	0.5014	0.5046	0.5055	0.5046	0.5026
MC(6)	0.4962	0.5027	0.5065	0.5077	0.5071	0.5052
MC(7)	0.4964	0.5037	0.5079	0.5095	0.5091	0.5073
MC(8)	0.4962	0.5040	0.5086	0.5106	0.5104	0.5087
MC(9)	0.4953	0.5036	0.5087	0.5109	0.5109	0.5093
MC(10)	0.4938	0.5025	0.5079	0.5103	0.5105	0.5091

Fig. 1. Counter-intuitive performance of Monte-Carlo agents in a Double Step Race 3 with 16 Black Holes. In this game, $MC(9)$, for example, defeats $MC(15)$ 51.09 % of the time.

The figure in the upper half of Fig. 1 shows the starting position of DSR-3 with 16 Black Holes. In the lower half, scores of $MC(i)$ vs $MC(k)$ are given in which both agents are equally likely to be first player. Vertically, i runs from 5 to 10, and horizontally, k from 11 to 16. The entry in the upper-left corner of the table, shows the likelihood that $MC(5)$ wins against $MC(11)$. It is 0.4955, and thus below 50 %. The interesting pairings here are the ones with entries above 0.5. In these, the "smaller" agent is more likely to win.

The two highlighted cells contain the table's maximum values. In these two cases, the Monte-Carlo agent that plays 9 random games for each candidate move will win more than 51 % of the time against an agent that plays 14 or 15 games, respectively. In the rest of this paper, the scores reported are exact values (here rounded to four decimals), determined with the help of a computer algebra program.

It is a non-trivial task to see that DSR-3 with 16 Black Holes is Monte-Carlo perfect. Figure 2 provides the necessary values for all possible positions. In this figure, each position of the game is encoded by a string of five symbols: distance of player White to his goal, presence of the white "hole jumper" (+ or -), distance of player Black to his goal, presence of the black "hole jumper", and a fifth symbol showing the player to move. The starting position (with White to move) is therefore given by 3+3+W. When White starts by jumping into a black hole, the resulting position is 3-3+B. Each node shows the likelihood White wins a random game from this position. Wins for White are denoted with a 1 and wins for Black with a 0.

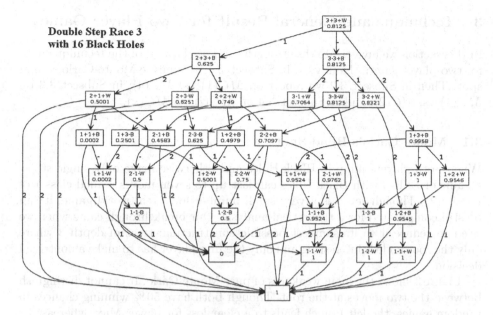

Fig. 2. On the Monte-Carlo perfection of DSR-3 with 16 Black Holes

The symbols beside the edges are the corresponding moves: 2 is a double step, 1 a single step, and – a jump into a black hole. Observe the interesting fact that in purely random games, jumping from the starting position into a black hole gives better winning chances than making a single step forward towards the target cell. However, because a double step is even better with respect to random continuations, this game is indeed Monte-Carlo perfect.

2.2 Length 3 Double Step Races with 47 Black Holes/Refined Double Step Races

Finding a game in which a lower-parameter pure-MC agent upsets a higher-parameter pure-MC agent more than half the time was our first success in this investigation. A search for the parameters in which such upsets were most likely led us to DSR-3 with 47 Black Holes (called class 2). Here $MC(11)$ wins 55 percent of its games against $MC(26)$. Interestingly (or disappointingly), continuing to increase the number of Black Holes (at least up to 5,000) did not produce any games in which smaller agents performed any better than this.

Next, we investigated variants of the game where the number of Black Holes was dependent on the positions of the pieces along the track called class 3). Though this led to findings with higher likelihoods of upset (even above 98 percent for games with 5,000 Black Holes), we saw no good chances for a theoretical proof with class 3.

3   Technique and General Result for Two-Player Games

In this section we investigate the strengths and weaknesses of the technique used for two-player games (Family 2). In Subsect. 3.1 we analyze Monte-Carlo's blind spot. Then in Subsect. 3.2 we report on $MC(1)$ vs. $MC(4)$; in Subsect. 3.3 on $MC(1)$ vs. $MC(2)$, and in Subsect. 3.4 on $MC(i)$ vs. $MC(k)$.

3.1   Monte-Carlo's Blind Spot

When both players jump in Black Holes and otherwise make only single steps, a game of DSR-3 with Black Holes can last up to seven moves (called class 1 of Family 2). Though seven is a very small number for a real world game, it can be still complicated for a theoretical analysis. Our breakthrough came when we tried to reduce it to its simplest possible construction: trees of depth 3 where only the root node is interesting and only the first player has to make a nontrivial decision.

Figure 3 shows the main weakness of pure Monte-Carlo. It cannot distinguish between the two moves at the root. Though both have 50 % winning chances in random games, the left branch leads to a clear loss for player Max, whereas the right branch gives him a simple win.

In the two-player games from the second class, the root will have three types of successors: a single node with perfect score 1 (this node makes the game Monte-Carlo perfect), **several** nodes of type B which lead to a trivial loss for player Max, and **many** nodes of type C which lead to a trivial win for player Max. In random games, however, type-B nodes have higher winning chances for Max than type-C nodes. For smaller i, $MC(i)$, in its sampling, will discover many type-C nodes to have perfect scores. For larger k, $MC(k)$ will find fewer 100 % wins of type C, but will still find many nodes of type B to be 100 %-likely wins. Applying Monte-Carlo's fair tiebreak rule, any of the candidate-moves with a perfect score will be selected with the same probability. Below we discuss one specific example in Subsect. 3.2. Then in Subsect. 3.3 we discuss class 3 of Family 2, and in Subsect. 3.4 the general case of $MC(i)$ vs. $MC(k)$, class 4 of Family 2.

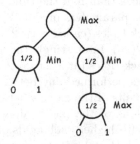

Fig. 3. Pure Monte-Carlo with a blind spot: left and right children look equally promising

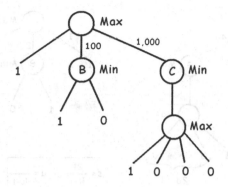

Fig. 4. A Monte-Carlo perfect game tree where $MC(1)$ is superior to $MC(4)$

3.2   69 Percent for $MC(1)$ vs. $MC(4)$

A medium-size example will explicate the method here. With the game tree in Fig. 4, we compare agents $MC(1)$ and $MC(4)$ in the role of player Max. In the root there are 100 moves to B-nodes and 1,000 moves to C-nodes.

When $MC(1)$ is Max, each C-node will receive a perfect score with probability 0.25, and each B-node with probability 0.5. There are therefore $100 \cdot 0.5 = 50$ expected type-B moves with a perfect score, $1,000 \cdot 0.25 = 250$ expected type C, and, of course, 1 perfect candidate of type 1. This means that the tiebreak procedure (where ties are broken between the nodes with perfect Max-scores) leads $MC(1)$ to a B-node with a probability of approximately $50/(1+50+250) \approx 0.166$.

In contrast, when $MC(4)$ is Max, each C-node will have a perfect score with probability $(1/4)^4 = 1/256$, and each B-node with probability $(1/2)^4 = 1/16$. So, on average $MC(4)$ will see $100 \cdot 1/16 = 1,600/256$ perfect candidates of type B, $1,000 \cdot 1/256 = 1,000/256$ perfect candidates of type C, and again the single perfect candidate on the left-hand side. With the fair-tiebreak procedure, $MC(4)$ chooses a B-node with probability of approximately $1,600/(1,600 + 1,000 + 256) \approx 0.560$. As $0.560 \gg 0.166$, $MC(4)$ has a much higher chance as player Max to stumble into a bad node of type B.

A more elaborate computation with exact probabilities shows that $MC(4)$ loses with a probability 0.5561 as Max whereas $MC(1)$ loses with only probability 0.1662 in that role. On average, $MC(1)$ beats $MC(4)$ in this game with probability $(0.5561 + 0.8338)/2 = 0.6949$.

3.3   Almost 100 Percent for $MC(1)$ vs. $MC(2)$

In a more abstract way, we now look at class 3 of Family 2 in which smaller-parametered Monte-Carlo agents perform arbitrarily well against larger. Figure 5a describes games in which $MC(1)$ beats $MC(2)$ and Fig. 5b generalizes this class to games in which $MC(i)$ handily beats $MC(j)$ given any $i < j$.

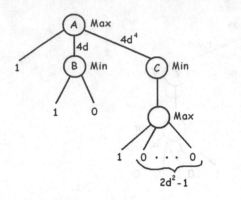
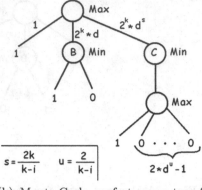

(a) Monte-Carlo perfect game trees for $MC(1)$ vs $MC(2)$ with the variable parameter d

(b) Monte-Carlo perfect game tree for $MC(i)$ vs $MC(k)$

Fig. 5. Two-player game trees in which pure Monte-Carlo agents with a lower parameter k beat those of a higher parameter with probability arbitrarily close to 1

As seen already, in Fig. 5a $MC(1)$ gets a perfect score in a sample of each C-node with probability $1/2d^2$, and each B-node with probability 0.5. On average, $MC(1)$ will see $2d$ perfect candidates of type B, $4d^4/2d^2 = 2d^2$ perfect candidates of type C, and the lone perfect candidate on the left. Under the tiebreak procedure, this leaves $MC(1)$ a chance of choosing a B-node with probability around $2d/(1 + 2d + 2d^2) \approx 1/d$.

In contrast, for $MC(2)$ each C-node will get a perfect Max-score (two wins in two random games) with probability $1/(2d^2)^2 = 1/4d^4$ and each B-node with probability $(1/2)^2 = 1/4$. So, on average, $MC(2)$ will see d B-nodes with a perfect score, $4d^4/4d^4 = 1$ perfect C-node candidates, and the single perfect 1-candidate. With tiebreaks, $MC(2)$ will choose a B-node with probability around $d/(1 + d + 1) = d/(d + 2) \approx 1 - 2/d$.

In this argument we have used expected values instead of the true distributions. One has to make sure that this is justified. This is easy to see in the case where the expected number of candidates is d or some higher power of d, because the variance is always linear in the numbers of candidates (and the standard deviation, accordingly, linear in the square root of the number of candidates). In this case, the distribution is strongly centered around its expected value.

It is more difficult when the expected number of candidates is constant (like in the case of perfect C-candidates for $MC(2)$, where it is 1). Here, the true values are approximately Poisson-distributed with parameter 1. The probability that the number of candidates equals n (here, p_n) is $(1/e) \cdot (1/n!)$ for all natural n, where e is Euler's constant. This gives

$$P(MC(2) \text{ reaches a C-node}) \approx \sum_{n=0}^{\infty} p_n \frac{n}{1+d+n}$$

$$\leq \sum_{n=0}^{\infty} p_n \frac{n}{d} \quad = \sum_{n=1}^{\infty} p_n \frac{n}{d} \quad = \sum_{n=1}^{\infty} \frac{1}{e} \cdot \frac{n}{n!} \cdot \frac{1}{d}$$

$$= \frac{1}{e \cdot d} \sum_{n=1}^{\infty} \frac{1}{(n-1)!} \quad = \frac{1}{e \cdot d} \cdot e \quad = \frac{1}{d}$$

3.4   The General Case: $MC(i)$ vs. $MC(k)$

The construction from Fig. 5a can be generalized to class 4 of Family 2, i.e., the MC-perfect game trees shown in Fig. 5b. In these, $MC(i)$ performs arbitrarily well against $MC(k)$ for any pair (i, k) with $i < k$.

In games from this class, B-nodes again result in a loss for player Max, while C nodes and type-1 nodes give him a win. With probability around $1 - 1/d$ agent $MC(k)$ will solve a tiebreak by going into a B, where $MC(i)$ will move to a C node with about the same probability. As a consequence, we can look at $\epsilon \approx 1/d$ and get the following result.

Theorem 1. *For each pair of integers (i, k) with $i < k$ and each $\epsilon, 0 < \epsilon < 1$, there exists a Monte-Carlo perfect two-player game tree of depth 3, where $MC(i)$ will win against $MC(k)$ with a probability of $1 - \epsilon$.*

4   Extreme Game Trees for Single Player Games

In this section we investigate the strengths and weaknesses of the techniques used for extreme game trees for single-player games (Family 3). In Subsect. 4.1 we discuss $MC(1)$ against $MC(2)$, and in Subsect. 4.2 $MC(i)$ against $MC(k)$.

4.1 $MC(1)$ Against $MC(2)$

The key idea in our single-player construction is to exploit the fact that pure Monte-Carlo game search works without memory. When $MC(k)$ moves from some node X to a successor Y, it does not remember which 1-leaves and branches this decision was based on. After the move is made it starts again from scratch.

In the tree of Fig. 6, the root has many B-successors and even more C-successors. C-nodes are the good successors, because from them it is trivial for Monte-Carlo to obtain a 1-leaf.

A random game starting at B ends in a 1-leaf with probability $1/d^2$, and a random game starting at C ends in a 1-leaf with probability $1/d^4$. As there are d^5 B-successors and d^8 C-successors of the root, $MC(1)$ will see on average $d^5/d^2 = d^3$ B-nodes with a full score, and $d^8/d^4 = d^4$ C-nodes with a full score. It will therefore select a B-node from the root with probability of about $1/d$. In comparison, $MC(2)$ will see on average $d^5/d^4 = d$ B-nodes with score 2 from

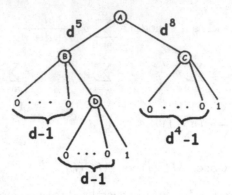

Fig. 6. Single-player game where $MC(1)$ scores about $1 - 1/d$ and $MC(2)$ only about $4/d$

its two random games, and only $d^8/d^8 = 1$ C-node with score 2. $MC(2)$ will therefore select a B-node with probability of about $d/(d + 1)$.

For our final step, we show that $MC(2)$ has some difficulty reaching a 1-leaf from B. A random game starting at D has probability $1/d$ to get the only 1-leaf. As a result, $MC(2)$ will get at least one 1-leaf in its two runs from D with a probability of less than $2/d$. If $MC(2)$ does not discover the win below D, it will go to any of the B-successors (including D) with probability $1/d$. In total, the probability that $MC(2)$ is successful is approximately

$$\frac{1}{d} + \left(1 - \frac{1}{d}\right) \cdot \left[\frac{2}{d} + \left(1 - \frac{2}{d}\right) \cdot \frac{1}{d}\right] < \frac{1}{d} + \left(1 - \frac{1}{d}\right) \cdot \left[\frac{2}{d} + \frac{1}{d}\right]$$

$$= \frac{1}{d} + \left(1 - \frac{1}{d}\right) \cdot \frac{3}{d}$$

$$< \frac{4}{d}$$

The first $1/d$ term in this equation comes from the case where $MC(2)$ goes from the root to a C-node. As a consequence we can formulate Theorem 2.

Theorem 2. *In the single-player game tree of Fig. 6, $MC(1)$ reaches a win with probability about $1 - 1/d$ and $MC(2)$ wins with probability less than $4/d$.*

4.2 $MC(i)$ against $MC(k)$

The construction of Fig. 6 is generalized in Fig. 7. Here we arrive at an analogous result in which $MC(i)$ outperforms $MC(k)$ for any $i < k$.

As an additional note, we observe that interpreting these single-player games as special two-player games gives examples where the performance of $MC(k)$ vs. perfect play does not grow monotonically with k.

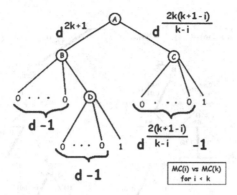

Fig. 7. Single-player game where $MC(i)$ scores about $1 - 1/d$ and $MC(k)$ only about k/d

5 Concrete Games with Anomalies

In the previous sections we described the construction of games in which Monte-Carlo agents with lower search parameters outperformed those with higher arbitrarily well. These games were designed to render our analysis tractable as we described the extent to which one agent outperformed another. In this section we explore game trees that were engineered to maximize the performance of one agent against another while simultaneously minimizing the size of the game. Subsects. 5.1 and 6.2 give two-player games (Family 2) that are much smaller than their counterparts in Sect. 3, while Subsects. 5.2, 6.1, and 6.2 explore the existence (and nonexistence) of smaller single-player games (Family 3).

5.1 A Small Two-Player Example

Below we discuss a small two-player example of Family 2, in which MC(1) defeats MC(100).

Let the *degree* of a game be the maximum degree of all its constituting positions. Figure 8, then, gives a two-player game tree constructed with the intention of maximizing the likelihood $MC(1)$ defeats $MC(100)$ while minimizing the game's degree.

As with the graphs in Sects. 3 and 4, we have annotated edges with their cardinality (here in parentheses) instead of drawing edges to each replicated position. For example, this game has 100 identical positions of type B and one node of type C. Each position is annotated with the likelihood that a random game from that position results in a win for Max (represented by the upward-facing triangles). Thick edges are winning moves and thin are losing. Sitting next to the edge's cardinality (with MC(1) on top and MC(100) on the bottom) is the likelihood that the corresponding Monte-Carlo agent chooses a node in that class given that it is at the parent.

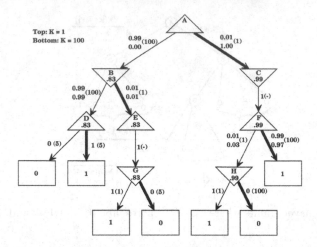

Fig. 8. A (relatively) small game in which $MC(1)$ defeats $MC(100)$ 97.6 % of the time

In this game, $MC(1)$ defeats $MC(100)$ 97.6 % of the time. When $MC(1)$ is Max it wins 97.9 % of its games. When $MC(100)$ is Max, it wins just 2.7 %.

This construction works by presenting Max with an initial position that is sufficiently difficult for $MC(1)$ to make a mistake almost always, but not so difficult that $MC(100)$ is fooled. To take advantage of $MC(1)$'s mistake, however, $MC(100)$ is forced to make a move that is beyond its ability to do so reliably. When $MC(100)$ is Max and makes the expected, correct move, it is faced with another decision to preserve its win that is also too difficult for it to make with high likelihood.

5.2   Small Instances of Single-Player Games

Below we discuss a small one-player example of Family in which $MC(11)$ outperforms $MC(2)$.

Built in the same style as Fig. 8 from the last subsection, Fig. 9 describes a pair of single-player games. Here, $MC(1)$ agents outperform $MC(2)$ agents to a greater degree than their counterparts in Fig. 6. Setting $d = 2$ in Fig. 6 gives a game in which $MC(2)$ actually scores slightly better than $MC(1)$. However, by adjusting the number of different types of nodes, we obtain the game shown in Fig. 9a in which $MC(1)$ does indeed outperform $MC(2)$. In this game, $MC(1)$ finds a win 89.003 % of the time (rounded to 5 figures) and $MC(2)$ with probability 87.200 % (for a difference of 1.803 %).

The game shown in Fig. 9b is the result of searching for single-player games of lowest possible degree which exhibit the non-monotonic convergence explored in this paper. We were unable to find any games at all with fewer than 104 nodes of type C in which $MC(1)$ plays better than $MC(2)$. In Fig. 9b, $MC(1)$ wins with probability 99.2746 % and $MC(2)$ with 99.2734 % (for a difference of 0.0012 %).

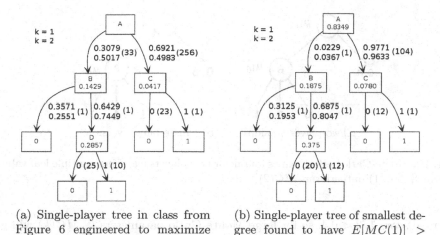

(a) Single-player tree in class from Figure 6 engineered to maximize $E[MC(1)] - E[MC(2)]$

(b) Single-player tree of smallest degree found to have $E[MC(1)] > E[MC(2)]$

Fig. 9. Relatively small single-player game trees in which $MC(1)$ outperforms $MC(2)$

6   Results over Very Small Game Trees

In this section we show the results over very small game trees. In Subsect. 6.1 we show the Family 3 results for small game trees: no anomalies. In Subsect. 6.2 we show multivalued trees with anomalies for Family 2 and Family 3 examples.

6.1   No Anomalies in Very Small Trees

The game trees in Subsect. 5.2, while smaller than the trees implied in Sect. 4, are still quite a bit larger than the ones in Subsect. 5.1. Inspired by the extremely small games that are about to be described in Subsect. 6.2, we began our search for single-player binary-valued game-trees that exhibit the property explored in this paper by exhaustively searching the space of small trees. However, after enumerating and examining every game with at most 26 nodes, we found no trees in which there exist a pair (i, k) with $1 \le i < k \le 10$ such that $MC(i)$ outperformed $MC(k)$. At around this point due to the exponential explosion in the number of possible trees an exhaustive search becomes prohibitive.

6.2   Very Small Multivalued Trees with Anomalies

To this point in this paper we have only discussed games in which leaves draw from only two values: win and loss. Early in our research, however, we found very simple game trees with leaves drawing from $\mathbb{R} \in [0, 1]$ where $MC(1)$ performs better than $MC(2)$. Figure 10 gives two such examples.

In the two-player version shown in Fig. 10a, the only optimal move for Max is to go to the left successor with value 0.4. Random games starting at B have an

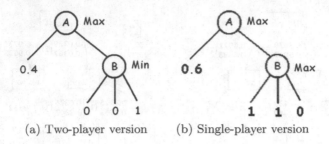

(a) Two-player version   (b) Single-player version

Fig. 10. Monte-Carlo perfect games for one or two players featuring multiple leaf values in which $MC(1)$ outperforms $MC(2)$

average score $1/3 < 0.4$, so for large Monte-Carlo parameters k, $MC(k)$ will go to the left with a probability of nearly 1. $MC(1)$, however, chooses the left with probability $2/3$ (the probability that a random game starting from B reaches a 0-leaf), and $MC(2)$ goes to the left with only probability $4/9$ (when in both random games starting from B a 0-leaf is found).

Analysis for the single-player version in Fig. 10b is analogous. In this case, the optimal move for Max (the only player in this game) is to move right to position B. Though for large k, $MC(k)$ will make this move almost every time, $MC(1)$ does so only $2/3$ of the time and $MC(2)$ only $4/9$ (when both samples find a 1-leaf).

7   Discussion and Open Problems

In this section we discuss six open problems. Each of the subsections (from Subsect. 7.1 to 7.6) contains an open problem.

Our investigation shows that agents running pure Monte-Carlo can exhibit pathological behavior when running a low number of simulations even when they perform asymptotically perfectly. Starting in the late 1970's with work by D.S. Nau [9] and D. Beal [10], pathological behavior has been observed and analyzed in a variety of models for two-player and single-player games.

Take, for example, a game tree in which only heuristic values are known at each leaf. When the actual values of each node and the errors in the evaluating heuristic are independently distributed, minimaxing in deeper trees often results in a higher probability of a mistaken root evaluation.

7.1   Connection and Comparison with Old Results of Game Tree Pathology

The models mentioned above are quite different from ours with the pure MC algorithm. However, a model somewhere in-between can be analyzed. From a result by Schrüfer [11], it follows that there exists, for instance, (many) binary game trees of large depth where

1. Very small error probabilities at the leaves lead to small error probability at the root
2. Medium-size error probabilities at the leaves lead to large error probability at the root
3. Large error probabilities at the leaves lead to medium-size error probability at the root.

In such trees, small error probabilities lead to better the best evaluations, but large error probabilities do better than medium-size error probabilities.

7.2   Anomalous Results for MCTS Strategies

Our example trees look rather special; they are constructed only for pure Monte-Carlo. One referee remarked that "nobody" would apply pure Monte-Carlo in practice, and that for instance the regret-minimizing algorithm UCB1 (UCB stands for "upper confidence bounds") by Auer, Cesa-Bianchi and Fischer ([12]) would have no problems with the trees from Fig. 5. However, a construction from J. Kahn [13] for a variant of MC in which at each node the same number of simulations are done (rather than the same number of simulations for each successor of that node) also gives examples where UCB1 performs non-monotonically with the number of simulations.

More generally, we are pretty sure that given any "reasonable," convergent game tree procedure P (be it MCTS, UCT or any other), we can find trees and run-time parameters n_1 and n_2 with $0 < n_1 < n_2$ such that $P(n_1)$ performs clearly better than $P(n_2)$.

Nevertheless, our results are weak in the following sense. They are only worst-case results. They say nothing about the performance of the procedure when amortized over a typical universe of instances. An analogous result is, for instance, well known from the Simplex algorithm(s) in linear programming: in most cases (particularly in practice) Simplex does well. However, degenerate instances (like the Klee-Minty polytopes from [14]) can be constructed for each detailed variant of Simplex. Our findings can thus not be compared with those as in [15] for typical trees.

7.3 $MC(k)$ vs. Perfect Play in Double Step Races

For DSR-3 with anywhere between 1 and 32 black holes we found one interesting side aspect: when examining the performance of $MC(k)$ vs perfect play $(MC(\infty))$, the winning probability of $MC(k)$ grows monotonically with k, for $1 \le k \le 18$. We are firmly convinced that this is true for all integers k, and might also be true for Double Step Races of length 3 with any number of Black Holes.

7.4   On Board Filling Games with Random Turn Order

Selection games ([16,17]) and board filling games [1] with random turn order have been proven to be Monte-Carlo perfect. We believe that games in these two classes cannot exhibit the anomalies.

7.5   Nonmonotonic Behavior of $MC(k)$ in Self-Play

In strict self-play, the score of first-player-$MC(k)$ vs second-player-$MC(k)$ does not necessarily converge to any value monotonically as k tends toward infinity. It would be interesting to find examples with wild performance curves. The classes from Sect. 4 could be potentially used as a starting point for generating such examples.

7.6   Anomalies in Real World Games

We are looking for interesting games from the real world with Monte-Carlo anomalies. These games do not need to be MC-perfect, but should be at least of a caliber where pure Monte-Carlo would give an interesting opponent for normal human players. A prominent candidate might be Connect Four.

Acknowledgments. This paper is dedicated to **Prof. Bernd Brügmann**. Twenty years ago in his report from 1993 [18] he introduced the approach of pure Monte-Carlo search to the game of Go and also created the name "Monte-Carlo Go". Brügmann now holds a chair in theoretical physics for gravitational theory. His special topic is the understanding of mergers of black holes. This motivated us to call the jump-outs in our Double Step Races "black holes".

Back in 2005, **Jörg Sameith** designed his beautiful tool *McRandom*, which allows one to design and test new board games with Monte-Carlo game search. With the help of *McRandom* we found the "DSR-3 with 16 Black Holes". Also the graphic in the upper part of Fig. 1 is that of *McRandom*. In a workshop on search methodologies in September 2012, **Soren Riis** asked several questions on our preliminary anomaly result. This motivated us to do the research presented in this paper. Participants of the computer-go mailing list gave helpful feedback; here special thanks go to **Jonas Kahn**, **Cameron Browne**, and **Stefan Kaitschick**. Thanks to **Matthias Beckmann** and **Michael Hartisch** for proofreading an earlier version, and to the three anonymous referees for their constructive criticism and helpful proposals. Finally, thanks to **Jaap van den Herik** for his proposals in giving the paper a much clearer structure.

References

1. Althöfer, I.: On games with random-turn order and Monte Carlo perfectness. ICGA J. **34**, 179–190 (2011)
2. Browne, C.: On the dangers of random playouts. ICGA J. **34**, 25–26 (2011)
3. Lorentz, R.J.: Amazons discover Monte-Carlo. In: van den Herik, H.J., Xu, X., Ma, Z., Winands, M.H.M. (eds.) CG 2008. LNCS, vol. 5131, pp. 13–24. Springer, Heidelberg (2008)
4. Coulom, R.: Efficient selectivity and backup operators in Monte-Carlo tree search. In: van den Herik, H.J., Ciancarini, P., Donkers, H.H.L.M.J. (eds.) CG 2006. LNCS, vol. 4630, pp. 72–83. Springer, Heidelberg (2007)
5. Kocsis, L., Szepesvári, C.: Bandit based Monte-Carlo planning. In: Fürnkranz, J., Scheffer, T., Spiliopoulou, M. (eds.) ECML 2006. LNCS (LNAI), vol. 4212, pp. 282–293. Springer, Heidelberg (2006)

6. Althöfer, I.: On the laziness of Monte Carlo game tree search in non-tight positions. Technical report, Friedrich Schiller University of Jena (2009). http://www.althofer. de/mc-laziness.pdf
7. Althöfer, I.: Game self-play with pure Monte Carlo: the basin structure. Technical report, Friedrich Schiller University of Jena (2010). http://www.althofer.de/ monte-carlo-basins-althoefer.pdf
8. Fischer, T.: Exakte Analyse von Heuristiken für kombinatorische Spiele. Ph.D. thesis, Fakultaet für Mathematik und Informatik, FSU Jena (2011). http://www. althofer.de/dissertation_thomas-fischer.pdf (in German language) Single step and double step races are covered in Chap. 2 of the dissertation
9. Nau, D.: Quality of decision versus depth of search on game trees. Ph.D. thesis, Duke University (1979)
10. Beal, D.: An analysis of minimax. In: Clarke, M. (ed.) Advances in Computer Chess, vol. 2, pp. 103–109. Edinburgh University Press, Edinburgh (1980)
11. Schrüfer, G.: Presence and absence of pathology on game trees. In: Beal, D. (ed.) Advances in Computer Chess, vol. 4, pp. 101–112. Elsevier, Amsterdam (1986)
12. Auer, P., Cesa-Bianchi, N., Fischer, P.: Finite-time analysis of the multiarmed bandit problem. Mach. Learn. **47**, 235–256 (2002)
13. Kahn, J.: Personal communication with I. Althöfer (2013)
14. Klee, V., Minty, G.J.: How good is the simplex algorithm? In: Shisha, O. (ed.) Proceedings of the Third Symposium on Inequalities, pp. 159–175. Academic Press, New York-London (1972)
15. Finnsson, H., Björnsson, Y.: Game-tree properties and MCTS performance. In: IJCAI'11 Workshop on General Intelligence in Game Playing Agents, pp. 23–30 (2011)
16. Peres, Y., Schramm, O., Sheffield, S., Wilson, D.B.: Random-turn hex and other selection games. Amer. Math. Mon. **114**, 373–387 (2007)
17. Wilson, D.B.: Hexamania - a computer program for playing random-turn hex, random-turn tripod, and the harmonic explorer game (2005). http://dbwilson. com/hex/
18. Brügmann, B.: Monte Carlo go (1993). Not published officially, but available at several places online including. http://www.althofer.de/Bruegmann-MonteCarloGo. pdf

Developments on Product Propagation

Abdallah Saffidine[✉] and Tristan Cazenave

LAMSADE, Université Paris-Dauphine, Paris, France
abdallah.saffidine@dauphine.fr, cazenave@lamsade.dauphine.fr

Abstract. Product Propagation (PP) is an algorithm to backup probabilistic evaluations for abstract two-player games. It was shown that
PP could solve Go problems as efficiently as Proof Number Search (PNS).
In this paper, we exhibit three domains where, for generic non-optimized
versions, PP performs better (see the nuances in the paper) than previously known algorithms for solving games. The compared approaches
include alpha-beta search, PNS, and Monte-Carlo Tree Search. We also
extend PP to deal with its memory consumption and to improve its
solving time.

1 Introduction

Product Propagation (PP) is a way to backup probabilistic information in a
two-player game-tree search [28]. It has been advocated as an alternative to
minimaxing that does not exhibit the minimax pathology [2,3,11,12,20].

PP was recently proposed as an algorithm to solve games, combining ideas
from Proof Number Search (PNS) and probabilistic reasoning [29]. In Stern's
paper, PP was found to be about as performant as PNS for capturing go problems.

We conduct a more extensive study of PP, comparing it to various other
paradigmatic solving algorithms and improving its memory consumption and its
solving time. Doing so, we hope to establish that PP is an important algorithm
for solving games that the game-search practician should know about. Indeed,
we exhibit multiple domains in which PP performs better than the other tested
game solving algorithms.

The baseline game-tree search algorithms that we use to establish PP's value
are Monte-Carlo Tree Search (MCTS) Solver [34] which was recently used to
solve the game of havannah on size 4 [9]; PNS [1,14,31]; and $\alpha\beta$ [21].

The next section deals with algorithms solving two-player games. The third
section is about the Product Propagation algorithm. The fourth section details
experimental results and the last section provides conclusion and further research.

2 Solving Two-Player Games

We assume a deterministic zero-sum two-player game with two outcomes and
sequential moves. Game-tree search algorithms have been proposed to address

H.J. van den Herik et al. (Eds.): CG 2013, LNCS 8427, pp. 100–109, 2014.
DOI: 10.1007/978-3-319-09165-5_9, © Springer International Publishing Switzerland 2014

games with multiple outcomes [7,22], multi-player games [19,25,30], non-deterministic games [10], and games with simultaneous moves [24].

A generic best-first search framework is presented in Algorithm 1. To instantiate this framework, one needs to specify a type of information to associate to nodes of the explored tree as well as functions to manipulate this type: info-term, init-leaf, select-child, update.

PNS is a best-first search algorithm which expands the explored game tree in the direction of most proving nodes, that is, parts of the tree which seem easier to prove or disprove.

Monte-Carlo Tree Search (MCTS) Solver is also a best-first search algorithm which can be cast in the mentioned framework. In MCTS [6], the information associated to nodes is in the form of sampling statistics, and bandit-based formulas [15] are used to guide the search. The sampling performed at a leaf node in MCTS can take the form of games played randomly until a terminal position, but it can also be the value of a heuristical evaluation function after a few random moves [17,33]. We denote the latter variant as MCTS-E.

3   Probability Search

In this section we describe the Product Propagation algorithm including the pseudocode (Sect. 3.1) and suggest practical improvements (Sect. 3.2).

3.1   The Product Propagation Algorithm

Product Propagation (PP) is a recently proposed algorithm to solve perfect information two-player two-outcome games based on an analogy with probabilities [29].

In PP, each node n is associated to a single number $PPN(n)$ (the probability propagation number for n) such that $PPN(n) \in [0,1]$. The PPN of a leaf corresponding to a *Max* win is 1 and the PPN of a *Max* loss is 0. $PPN(n)$ can intuitively be understood as the likelihood of n being a *Max* win given the partially explored game tree. With this interpretation in mind, natural update rules can be proposed. If n is an internal *Min* node, then it is a win for *Max* if and only if all children are a win for *Max* themselves. Thus, the probability that n is a win is the joint probability that all children are a win. If we assume all children are independent, then we obtain that the PPN of n is the product of the PPN of the children for *Min* nodes. A similar line of reasoning leads to the formula for *Max* nodes. To define the PPN of a non-terminal leaf l, the simplest is to assume no information is available and initiate $PPN(l)$ to $\frac{1}{2}$. These principles allow to induce a PPN for every explored node in the game tree and are summed up in Table 1.

Note that this explanation is just a loose interpretation of $PPN(n)$ and not a formal justification. Indeed, the independence assumption does not hold in practice, and in concrete games n is either a win or a loss for *Max* but it is not a random event. Still, the independence assumption is used because it is simple and the algorithm works well even though the assumption is usually wrong.

```
bfs(state q, player m)
    r ← new node with label m
    r.info ← init-leaf(r)
    n ← r
    while r is not solved do
        while n is not a leaf do
            n ← select-child(n)
        extend(n)
        n ← backpropagate(n)
    return r

extend(node n)
    switch on the label of n do
        case terminal
            n.info ← info-term(n)
        case max
            foreach q' in {q', q →ᵃ q'} do
                n' ← new node with label min
                n'.info ← init-leaf(n')
                Add n' as a child of n
        case min
            foreach q' in {q', q →ᵃ q'} do
                n' ← new node with label max
                n'.info ← init-leaf(n')
                Add n' as a child of n

backpropagate(node n)
    new_info ← update(n)
    if new_info = n.info ∨ n = r then  return n

    else
        n.info ← new_info
        return backpropagate(n.parent)
```

Algorithm 1. Pseudo-code for a best-first search algorithm.

Table 1. Initial values for leaf and internal nodes in PP. C denote the set of children.

	Node label	PPN
info-term	*Max* wins	1
	Max loses	0
init-leaf		$\frac{1}{2}$
update	*Max*	$1 - \prod_C (1 - \text{PPN})$
	Min	$\prod_C \text{PPN}$

To be able to use the generic best-first search (bfs) framework, we still need to specify which leaf of the tree is to be expanded. The most straightforward approach is to select the child maximizing PPN when at a *Max* node, and to select the child minimizing PPN when at a *Min* node, as shown in Table 2.

Table 2. Selection policy for PP. C denotes the set of children.

Node label	Chosen child
Max	$\arg\max_C \text{PPN}$
Min	$\arg\min_C \text{PPN}$

3.2 Practical Improvements

The *mobility* heuristic provides a better initialization for non-terminal leaves. Instead of setting PPN to $1/2$ as described in Table 1, we use an initial value that depends on the number of legal moves and on the type of node. Let c be the number of legal moves at a leaf, the PPN of which we want to initialize. If the leaf is a *Max*-node, then we set $\text{PPN} = 1 - 1/2^c$. If the leaf is a *Min*-node, then we set $\text{PPN} = 1/2^c$.

In the description of best first search algorithms given in Algorithm 1, we see that new nodes are added to the memory after each iteration of the main loop in **bfs**. Thus, if the **init-leaf** procedure is very fast then the resulting algorithm will fill the memory very quickly. Earlier work on PNS provides inspiration to address this problem [14]. For instance, Kishimoto proposed to turn PP into a depth-first search algorithm with a technique similar to the used in *dfpn*.[1]

Alternatively, it is possible to adapt the PN^2 ideas to develop a PP^2 algorithm. In PP^2, instead of initializing directly a non-terminal leaf, we call the PP algorithm on the position corresponding to that leaf with a bound on the number of nodes. The bound on the number of nodes allowed in the sub-search is set to the number of nodes that have been created so far in the main search. After a sub-search is over, the children of the root of that search are added to the tree of the main search. Thus, the PPN associated to these newly added nodes is based on information gathered in the sub-search, rather than based only on an initialization heuristic.

4   Experimental Results

While the performance of PP as a solver has matched that of PNS in Go [29], it has proven to be disappointing in shogi (See footnote 1). Below we exhibit three domains (Y, Domineering, NoGo) where the PP search paradigm outperforms more classical algorithms.

---

[1] Akihiro Kishimoto, personal communication.

In the following sets of experiments, we do not use any domain specific knowledge. We are aware that the use of such techniques would improve the solving ability of all our programs. Nevertheless, we believe that showing that a generic and non-optimized implementation of PP performs better than generic and non-optimized implementations of PNS, MCTS, or $\alpha\beta$ in a variety of domains provides good reason to think that the ideas underlying PP are of importance in game solving.

We have described a mobility heuristic for PP variants in Sect. 3.2. We also use the classical mobility heuristic for PNS variants. That is, if c is the number of legal moves at a non-terminal leaf to be initialized, then instead of setting the proof and disproof numbers to 1 and -1 respectively, we set them to 1 and c if the leaf is a *max*-node or to c and -1 if the leaf is a *min*-node.

All variants of PNS, PP, and MCTS were implemented with the best-first scheme described in Sect. 2. For PN^2 and PP^2, only the number of nodes in the main search is displayed.

4.1   The Game of Y

The game of Y was discovered independently by Claude Shannon in the 1950s, and in 1970 by Schensted and Titus [26]. It is played on a triangular board with a hexagonal paving. Players take turns adding one stone of their color on empty cells of the board. A player wins when they succeed in connecting all three edges with a single connected group of stones of their color. Just as hex, Y enjoys the no-draw property.

The current best evaluation function for Y is the *reduction evaluation function* [32]. This evaluation function naturally takes values in $[0, 1]$ with 0 (resp. 1) corresponding to a *Min* (resp. *Max*) win.

PNS with the mobility initialization could not solve any position in less than 3 min in a preliminary set of about 50 positions. As a result we did not include this solver in our experiment with a larger set of positions. The experiments on y was carried out as follows. We generated 77,012 opening positions on a board of size 6. We then ran PP using the reduction evaluation function, MCTS using playouts with a random policy, and a variant of MCTS using the same reduction evaluation instead of random playouts (MCTS-E). For each solver, we recorded the total number of positions solved within 60 s. Then, for each solving algorithm, we computed the number of positions among those 77,012 which were solved faster by this solver than by the two other solvers, as well as the number of positions which needed fewer iterations of the algorithm to be solved. The results are presented in Table 3.

We see that the PP algorithm was able to solve the highest number of positions, 77,010 positions out of 77,012 could be solved within 60 s. We also note that for a very large proportion of positions (68,477), PP is the fastest algorithm. However, MCTS needs fewer iterations than the other two algorithms on 35,444 positions. A possible interpretation of these results is that although the iterations of MCTS are a bit more informative than the iterations of PP, they take much longer. As a result, PP is better suited to situations where time is the most

Table 3. Number of positions solved by each algorithm and number of positions on which each algorithm was performing best.

	PP	MCTS	MCTS-E
Positions solved	77,010	76,434	69,298
Solved fastest	68,477	3,645	4,878
Fewest iterations	22,621	35,444	18,942

important constraint, while MCTS is more appropriate when memory efficiency is a bottleneck. Note that if we discard MCTS-E results, then 72,830 positions are solved fastest by PP, 4180 positions are solved fastest by MCTS, 30,719 positions need fewest iterations to be solved by PP, and 46,291 need fewest iterations by MCTS.

Figure 1 displays some of these results graphically. We sampled about 150 positions of various difficulty from the set of 77,012 y positions, and plotted the time needed to solve such positions by each algorithm against the time needed by PP. We see that positions that are easy for PP are likely to be easy for both MCTS solvers, while positions hard for PP are likely to be hard for both other solvers as well.

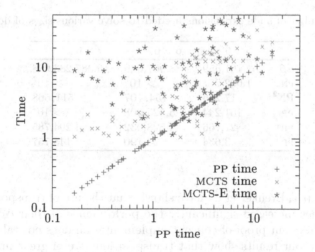

Fig. 1. Time needed to solve various opening positions in the game of y.

4.2   Domineering

Domineering is played on a rectangular board. The first player places a vertical 2×1 rectangle anywhere on the board. The second player places an horizontal 2×1 rectangle, and the games continues like that until a player has no legal moves. The first player that has no legal moves has lost.

Domineering has already been studied in previous work by game search specialists as well as combinatorial game theorists [4,16].[2] While these papers focusing on domineering obtain solutions for relatively large boards, we have kept ourselves to a naive implementation of both the game rules and the algorithms. In particular, we do neither perform any symmetry detection nor make use of combinatorial game theory techniques such as decomposition into subgames.

We present results for the following algorithms: $\alpha\beta$, PNS with Transpositions (PNT) [27], PN^2 [5], PP, PP with Transpositions (PPT) and PP^2. The PNS algorithm could not find a single solution within 10^7 node expansions when transpositions were not detected. Thus it is left out.

For PNS variants the standard mobility heuristic is used to compute the proof numbers and the disproof numbers at non-solved leaves. For PP variants, we used the mobility heuristic as described in Sect. 3.2.

Tables 4 and 5 give the number of nodes and times for different algorithms solving Domineering (on 5×6, 6×6, and 7×6 boards). $\alpha\beta$ is enhanced with transposition tables, killer moves, the history heuristic and an evaluation function. We can see that on the smallest 5×6 board that PPT gives the best results. On the larger 6×6 board PPT is the best algorithm by far. On the largest 7×6 board, several algorithms run out of memory, and the best algorithm remains PPT which outperforms both $\alpha\beta$ and PN^2.

Table 4. Number of node expansions needed to solve various sizes of domineering.

	5×6	6×6	7×6
$\alpha\beta$	701,559	38,907,049	6,387,283,988
PNT	1,002,277	$> 10^7$	$> 10^7$
PN^2	17,236	$> 154,107$	$> 511,568$
PP	101,244	5,525,608	$> 10^7$
PPT	27,766	528,032	4,294,785
PP^2	3,634	24,190	145,757

In their paper, Breuker *et al.* have shown that the use of transposition tables and symmetries increased significantly the performance of their $\alpha\beta$ implementation [4]. While, our proof-of-concept implementation does not take advantage of symmetries, our results show that transpositions are of great importance in the PP paradigm as well.

4.3   nogo

nogo is the misere version of the game of go. It was presented in the BIRS 2011 workshop on combinatorial game theory [8].[3] The first player to capture has lost.

---

[2] Some results can also be found on http://www.personeel.unimaas.nl/uiterwijk/ Domineering_results.html.

[3] http://www.birs.ca/events/2011/5-day-workshops/11w5073

Table 5. Time (s) needed to solve various sizes of domineering.

	5×6	6×6	7×6
$\alpha\beta$	0.87	40.68	5,656
PNT	5.92		
PN^2	78.7	>10,660	>153,000
PP	0.24	20.1	>35.84
PPT	0.17	5.33	55.13
PP^2	0.22	15.5	320.3

We present results for the following algorithms: $\alpha\beta$, PNT [27], PN^2 [5], PP, PPT and PP^2. Again, the PNS algorithm could not find a single solution within 10^7 node expansion and is left out.

For standard board sizes such as 4×4 or 5×4, $\alpha\beta$ gives the best results among the algorithms we study in this paper. We have noticed that for $N \times 1$ boards for $N \geq 18$, PPT becomes competitive. Results for a few board sizes are given in Table 6 for the number of nodes and in Table 7 for the times.

Table 6. Number of node expansions needed to solve various sizes of nogo.

	4×4	18×1	20×1	22×1
$\alpha\beta$	17,194,590	4,444,384	154,006,001	3,133,818,285
PNT	3,575,076	2,015,179	>10^7	>10^7
PN^2	77,010	> 22,679	> 29,098	
PP	>10^7	864,951	6,173,393	>10^7
PPT	2,319,816	98,991	389,119	2,814,553
PP^2		14,246		

Table 7. Time (s) needed to solve various sizes of nogo.

	4×4	18×1	20×1	22×1
$\alpha\beta$	33.05	10.43	361.0	7,564
PNT	436.6	144.2	> 809	
PN^2	27,519	> 3,607	> 4,583	
PP	> 338.84	21.39	156.3	> 307.55
PPT	396.36	9.46	46.3	446.58
PP^2		109.7		

5   Conclusions and Future Research

In this paper, we have presented how to use Product Propagation (PP) in order to solve abstract two-player games. We extended PP so as to handle transpositions and to reduce memory consumption with the PP^2 algorithm. For two of the games that have been tested (i.e., y, domineering), we found that our extensions of PP are better in solving games than the other solving algorithms. For nogo, PP variants outperform PNS variants on all tested sizes, and PP does better than $\alpha\beta$ on some sizes but $\alpha\beta$ is better on standard sizes.

Being a best-first search algorithm, PP is quite related to PNS and MCTS. As such, it seems natural to try and adapt ideas that proved successful for these algorithms to the Product Propagation paradigm. For instance, while PNS and PP are originally designed for two-outcome games, future work could adapt the ideas underlying Multiple-Outcome PNS [22] to turn PP into an algorithm addressing more general games. Adapting more elaborate schemes for transpositions could also prove interesting [13,18,23].

References

1. Allis, L.V., van der Meulen, M., van den Herik, H.J.: Proof-number search. Artif. Intell. **66**(1), 91–124 (1994)
2. Baum, E.B., Smith, W.D.: Best play for imperfect players and game tree search. Technical report, NEC Research Institute (1993)
3. Baum, E.B., Smith, W.D.: A Bayesian approach to relevance in game playing. Artif. Intell. **97**(1–2), 195–242 (1997)
4. Breuker, D.M., Uiterwijk, J.W.H.M., van den Herik, H.J.: Solving 8×8 domineering. Theor. Comput. Sci. **230**(1–2), 195–206 (2000)
5. Breuker, D.M.: Memory versus search in games. Ph.D. thesis, Universiteit Maastricht (1998)
6. Browne, C., Powley, E., Whitehouse, D., Lucas, S., Cowling, P., Rohlfshagen, P., Tavener, S., Perez, D., Samothrakis, S., Colton, S.: A survey of Monte Carlo tree search methods. IEEE Trans. Comput. Intell. AI Games **4**(1), 1–43 (2012)
7. Cazenave, T., Saffidine, A.: Score bounded Monte-Carlo tree search. In: van den Herik, H.J., Iida, H., Plaat, A. (eds.) CG 2010. LNCS, vol. 6515, pp. 93–104. Springer, Heidelberg (2011)
8. Chou, C.-W., Teytaud, O., Yen, S.-J.: Revisiting Monte-Carlo tree search on a normal form game: NoGo. In: Di Chio, C., et al. (eds.) EvoApplications 2011, Part I. LNCS, vol. 6624, pp. 73–82. Springer, Heidelberg (2011)
9. Ewalds, T.: Playing and solving Havannah. Master's thesis, University of Alberta (2012)
10. Hauk, T., Buro, M., Schaeffer, J.: Rediscovering *-MINIMAX search. In: van den Herik, H.J., Björnsson, Y., Netanyahu, N.S. (eds.) CG 2004. LNCS, vol. 3846, pp. 35–50. Springer, Heidelberg (2006)
11. Horacek, H.: Towards understanding conceptual differences between minimaxing and product-propagation. In: 14th European Conference on Artificial Intelligence (ECAI), pp. 604–608 (2000)
12. Horacek, H., Kaindl, H.: An analysis of decision quality of minimaxing vs. product propagation. In: Proceedings of the 2009 IEEE International Conference on Systems, Man and Cybernetics, SMC'09, Piscataway, NJ, USA, pp. 2568–2574. IEEE Press (2009)
13. Kishimoto, A., Müller, M.: A solution to the GHI problem for depth-first proof-number search. Inf. Sci. **175**(4), 296–314 (2005)

14. Kishimoto, A., Winands, M.H.M., Müller, M., Saito, J.-T.: Game-tree search using proof numbers: the first twenty years. ICGA J. **35**(3), 131–156 (2012)
15. Kocsis, L., Szepesvári, C.: Bandit based Monte-Carlo planning. In: Fürnkranz, J., Scheffer, T., Spiliopoulou, M. (eds.) ECML 2006. LNCS (LNAI), vol. 4212, pp. 282–293. Springer, Heidelberg (2006)
16. Lachmann, M., Moore, C., Rapaport, I.: Who Wins Domineering on Rectangular Boards, vol. 42. Cambridge University Press, Cambridge (2002)
17. Lorentz, R.J.: Amazons discover Monte-Carlo. In: van den Herik, H.J., Xu, X., Ma, Z., Winands, M.H.M. (eds.) CG 2008. LNCS, vol. 5131, pp. 13–24. Springer, Heidelberg (2008)
18. Müller, M.: Proof-set search. In: Schaeffer, J., Müller, M., Björnsson, Y. (eds.) CG 2002. LNCS, vol. 2883, pp. 88–107. Springer, Heidelberg (2003)
19. Nijssen, J.A.M., Winands, M.H.M.: An overview of search techniques in multi-player games. In: Computer Games Workshop at ECAI 2012, pp. 50–61 (2012)
20. Pearl, J.: On the nature of pathology in game searching. Artif. Intell. **20**(4), 427–453 (1983)
21. Russell, S.J., Norvig, P.: Artificial Intelligence - A Modern Approach, 3rd edn. Pearson Education, Upper Saddle River (2010)
22. Saffidine, A., Cazenave, T.: Multiple-outcome proof number search. In: De Raedt, L., Bessiere, C., Dubois, D., Doherty, P., Frasconi, P., Heintz, F., Lucas, P. (eds.) 20th European Conference on Artificial Intelligence (ECAI). Frontiers in Artificial Intelligence and Applications, Montpellier, France, vol. 242, pp. 708–713. IOS Press, August 2012
23. Saffidine, A., Cazenave, T., Méhat, J.: UCD: Upper Confidence bound for rooted Directed acyclic graphs. Knowl. Based Syst. **34**, 26–33 (2011)
24. Saffidine, A., Finnsson, H., Buro, M.: Alpha-beta pruning for games with simultaneous moves. In: Hoffmann, J., Selman, B. (eds.) 26th AAAI Conference on Artificial Intelligence (AAAI), Toronto, Canada, pp. 556–562. AAAI Press, July 2012
25. Schadd, M.P.D., Winands, M.H.M.: Best reply search for multiplayer games. IEEE Trans. Comput. Intell. AI Games **3**(1), 57–66 (2011)
26. Schensted, C., Titus, C.: Mudcrack Y & Poly-Y. Neo Press (1975)
27. Schijf, M., Victor Allis, L., Uiterwijk, J.W.H.M.: Proof-number search and transpositions. ICCA J. **17**(2), 63–74 (1994)
28. Slagle, J.R., Bursky, P.: Experiments with a multipurpose, theorem-proving heuristic program. J. ACM **15**(1), 85–99 (1968)
29. Stern, D., Herbrich, R., Graepel, T.: Learning to solve game trees. In: 24th International Conference on Machine Learning, ICML '07, pp. 839–846. ACM, New York (2007)
30. Sturtevant, N.R.: A comparison of algorithms for multi-player games. In: Computer and Games (2002)
31. van den Herik, H.J., Winands, M.H.M.: Proof-number search and its variants. In: Tizhoosh, H.R., Ventresca, M. (eds.) Oppos. Concepts in Comp. Intel., SCI, vol. 155, pp. 91–118. Springer, Heidelberg (2008)
32. van Rijswijck, J.: Search and evaluation in Hex. Technical report, University of Alberta (2002)
33. Winands, M.H.M., Björnsson, Y., Saito, J.-T.: Monte Carlo tree search in lines of action. IEEE Trans. Comput. Intell. AI Games **2**(4), 239–250 (2010)
34. Winands, M.H.M., Björnsson, Y., Saito, J.-T.: Monte-Carlo tree search solver. In: van den Herik, H.J., Xu, X., Ma, Z., Winands, M.H.M. (eds.) CG 2008. LNCS, vol. 5131, pp. 25–36. Springer, Heidelberg (2008)

Solution Techniques for Quantified Linear Programs and the Links to Gaming

Ulf Lorenz, Thomas Opfer[(✉)], and Jan Wolf

Fluid Systems Technology, Technische Universität Darmstadt,
Darmstadt, Germany
Thomas.Opfer@fst.tu-darmstadt.de

Abstract. Quantified linear programs (QLPs) are linear programs (LPs) with variables being either existentially or universally quantified. QLPs are two-person zero-sum games between an existential and a universal player on the one side, and convex multistage decision problems on the other side. Solutions of feasible QLPs are so called winning strategies for the existential player that specify how to react on moves – well-thought fixations of universally quantified variables – of the universal player to be sure to win the game. To find a certain best strategy among different winning strategies, we propose the extension of the QLP decision problem by an objective function. To solve the resulting QLP optimization problem, we exploit the problem's hybrid nature and combine linear programming techniques with solution techniques from game-tree search. As a result, we present an extension of the Nested Benders Decomposition algorithm by the $\alpha\beta$-algorithm and its heuristical move-ordering as used in game-tree search to solve minimax trees. The applicability of our method to both QLPs and models of PSPACE-complete games such as Connect6 is examined in an experimental evaluation.

1 Introduction

In the last 40 years, algorithmic achievements such as the introduction of the Alpha-Beta-Algorithm strengthened game-tree search remarkably [4,7,12,15]. Nowadays it is almost impossible for human chess masters to beat high-level chess computers. Proof Number Search has shown to be an effective algorithm for solving games [1,16,18]. Interestingly, there is a strong relation between these Artificial Intelligence techniques and effective treatment of uncertainty in Operations Research applications.

For many decades, a large amount of practical problems have been modeled as linear or mixed-integer linear programs (MIPs), which are well understood and can be solved quite effectively. However, there is a need for planning and deciding under uncertainty, as companies observe an increasing danger of disruptions, which prevent them from acting as planned. One reason is that input data for a given problem is often assumed to be deterministic and exactly known when decisions have to be made, but in reality they are often afflicted with some

H.J. van den Herik et al. (Eds.): CG 2013, LNCS 8427, pp. 110–124, 2014.
DOI: 10.1007/978-3-319-09165-5_10, © Springer International Publishing Switzerland 2014

kinds of uncertainties. Examples are flight and travel times, throughput-time, or arrival times of externally produced goods.

Uncertainty, often pushes the complexity of problems that are in P or NP, to the complexity class PSPACE. Also many optimization problems under uncertainty are PSPACE-complete [10] and therefore, NP-complete integer programs are not suitable to model these problems anymore. Relatively unexplored are the abilities of linear programming extensions for PSPACE-complete problems. In this context, Subramani introduced the notion of quantified linear programs (QLPs) [13,14]. While it is known that quantified linear integer programs (QIPs) are PSPACE-complete, the exact complexity class of their QLP-Relaxations is unknown in general. It turned out that quantified mixed-integer programming is a suitable modeling language for both optimization under uncertainty and games [5,6]. The idea of our research is to explore the abilities of linear programming techniques when being applied to PSPACE-complete problems and their combination with techniques from other fields.

In this paper we show how the problem's hybrid nature of being both a two-person zero-sum game on the one hand, and being a convex multistage decision problem on the other hand, can be utilized to combine linear programming techniques with techniques from game-tree search. To the best of our knowledge, a combination of techniques from these two fields has not been done before. Solutions of feasible QLPs are so called winning strategies for the existential player that specify how to react on moves - well-thought fixations of universally quantified variables - of the universal player to be sure to win the game. However, if there are several winning strategies, one might wish to find a certain (the best) one with respect to some kind of measure. We therefore propose an extension of the QLP decision problem by the addition of a linear objective function, which tries to minimize the objective function with respect to the maximum possible loss that can result from the universal player's possible decisions. To solve the resulting QLP optimization problem, we propose an extension of the Nested Benders Decomposition algorithm as presented in [5]. Solving a QLP with this algorithm can be interpreted as solving a tree of linear programs by passing information among nodes of the tree, and for the special case of QLPs with an objective function, the way the information items are passed is similar to the minimax principle as it is known from game-tree search. This allows the integration of $\alpha\beta$-cuts in combination with move-ordering, as used in the $\alpha\beta$-algorithm. The applicability of this approach is examined in an experimental evaluation, where we solve a set of QLPs that were generated from the well-known Netlib test set. Revealing that quantified programming is not only itself a game but can also be used to model and solve conventional games, we also model the Connect6 game as quantified program.

The rest of this paper is organized as follows. In Sect. 2, we formally describe the QLP optimization problem, followed by an explanation of the Nested Benders Decomposition approach in Sect. 3. Section 4 introduces the concepts of game-tree search and afterwards. In Sect. 5, we show how these techniques can be embedded into our existing algorithmic framework. We proceed with an experimental evaluation in Sect. 6. In Sect. 7, we give an outlook how to apply our method to PSPACE complete games. We end up with a conclusion in Sect. 8.

2   The Problem Statement: Quantified Linear Programs (QLPs)

Within this paper, we intend to concentrate on quantified linear programs (QLP), as they were introduced in [13,14], and in-depth analyzed in [5,9]. In contrast to traditional linear programs where all variables are implicitly existentially quantified, QLPs are linear programs with the variables being either existentially or universally quantified.

Definition 1 (Quantified Linear Program). *Let there be a vector of n variables $x = (x_1, \ldots, x_n)^T \in \mathbb{Q}^n$, lower and upper bounds $l \in \mathbb{Z}^n$ and $u \in \mathbb{Z}^n$ with $l_i \leq x_i \leq u_i$, a coefficient matrix $A \in \mathbb{Q}^{m \times n}$, a right-hand side vector $b \in \mathbb{Q}^m$ and a vector of quantifiers $Q = (Q_1, \ldots, Q_n)^T \in \{\forall, \exists\}^n$. Let the term $Q \circ x \in [l, u]$ with the component wise binding operator $\circ$ denote the quantification vector $(Q_1 x_1 \in [l_1, u_1], \ldots, Q_n x_n \in [l_n, u_n])^T$ such that every quantifier Q_i binds the variable x_i ranging over the interval $[l_i, u_i]$. We call (Q, l, u, A, b) with*

$$Q \circ x \in [l, u] : Ax \leq b \qquad \text{(QLP)}$$

a quantified linear program (QLP).

We denote the quantification vector $Q \circ x \in [l, u]$ as quantification sequence $Q_1 x_1 \in [l_1, u_1] \ldots Q_n x_n \in [l_n, u_n]$. In a similar manner, we denote Q as a quantifier sequence $Q_1 \ldots Q_n$ and x as a variable sequence $x_1 \ldots x_n$. Each maximal consecutive subsequence of Q consisting of identical quantifiers is called a *quantifier block* – the corresponding subsequence of x is called a *variable block*. The total number of blocks less one is the number of *quantifier changes*. In contrast to traditional linear programs, the order or the variables in the quantification vector is of particular importance.

Each QLP instance is a two-person zero-sum game between an *existential player* setting the $\exists$-variables and a *universal player* setting the $\forall$-variables. Each fixed vector $x \in [l, u]$, that is, when the existential player has fixed the existential variables and the universal player has fixed the universal variables, is called a *game*. If x satisfies the linear program $Ax \leq b$, we say *the existential player wins*, otherwise *the universal player wins*. The variables are set in consecutive order according to the quantification sequence. Consequently, we say that a player makes the move $x_k = z$, if he fixes the variable x_k to the value z. At each such move, the corresponding player knows the settings of $x_1, \ldots, x_{k-1}$ before taking his decision x_k. In the context of answering the question whether the existential player can certainly win the game, we use the term *policy*.

Definition 2 (Policy). *Given a QLP (Q, l, u, A, b) with $Q \circ x \in [l, u] : Ax \leq b$. An algorithm that fixes all existential variables x_i with the knowledge, how $x_1, \ldots, x_{i-1}$ have been set before, is called a policy.*

A policy can be represented as a set of computable functions of the form $x_i = f_i(x_1, \ldots, x_{i-1})$ for all existentially quantified variables x_i. A policy is called a *winning policy* if these functions ensure that the existential player wins all games that can result from this policy, independently of the universal player's moves.

Definition 3 (QLP Decision Problems). *Given a QLP, the decision problem "Is there a winning policy for the existential player?" is called the QLP Decision Problem.*

It has been shown that the QLP problem with only one quantifier change is either in P (when the quantification begins with existential quantifiers and ends with universal ones) or coNP-complete (when the quantification begins with universal quantifiers and ends with existential ones) [14]. In [9] it was shown that the solution space of a QLP with n variables forms a polytope in $\mathbb{R}^n$, which is included in the polytope induced by the constraint set $Ax \leq b$ as shown in Fig. 1. It was furthermore shown that it suffices to inspect the bounds of the universal quantified variables in order to check whether a *winning policy* does exist (cf. [14] and with a completely different proof in [9]).

Example 1. *The QLP*

$$\exists x_1 \in [1,6] \ \forall x_2 \in [1,2] : x_1 + x_2 \leq 6 \ \wedge \ x_2 - x_1 \leq 0$$

has the following graphical representation of bounding box (dashed lines) and constraints (solid lines).

We say a solution to this problem is a move for the existential player such that he wins the game regardless of the universal player's reaction, or rather, the set of games $(x_1, x_2)^T$ which can result from the existential player's decision (black line segment in Fig. 1), e.g. the output of a winning policy. The set of all solutions, i.e., the set of 'existential sticks' fitting in the specified trapezoid, is called the solution space *(filled rectangle). In this example, we see that it indeed suffices to analyze discrete points (filled dots) to find a solution. Note that the order in the quantification sequence is crucial.*

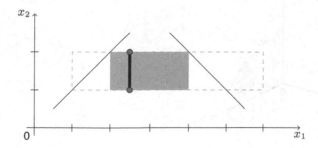

Fig. 1. Polyhedral QLP solution space

The fact that it suffices to check the bounding values of the universally quantified variables in order to answer the question whether the existential player can certainly win the game, can be exploited in terms of asking whether a *winning strategy* for the existential player does exist.

Definition 4 (Strategy). *A strategy $S = (V, E, c)$ is an edge-labeled finite arborescence with a set of nodes $V = V_\exists \,\dot\cup\, V_\forall$, a set of edges E and a vector of edge labels $c \in \mathbb{Q}^{|E|}$. Each level of the tree consists either of only nodes from $V_\exists$ or only of nodes from $V_\forall$, with the root node at level 0 being from $V_\exists$. The i-th variable of the QLP is represented by the inner nodes at depth $i - 1$. Each edge connects a node in some level i to a node in level $i + 1$. Outgoing edges represent moves of the player at the current node, the corresponding edge labels encode the variable allocations of the move. Each node $v_\exists \in V_\exists$ has exactly one child, and each node $v_\forall \in V_\forall$ has two children, with the edge labels being the corresponding upper lower and upper bounds.*

A path from the root to a leaf represents a game of the QLP and the sequence of edge labels encodes its moves. A strategy is called a *winning strategy* if all paths from the root node to a leaf represent a vector x such that $Ax \leq b$. This terminology is also very similarly used in game-tree search [11].

Example 2. *The QLP*

$$\exists x_1 \in [0,1] \; \forall x_2 \in [0,1] \; \exists x_3 \in [0,1] :$$

$$\begin{pmatrix} 0 & -1 & -1 \\ -1 & 1 & 1 \\ 2 & 2 & 0 \end{pmatrix} \begin{pmatrix} x_1 \\ x_2 \\ x_3 \end{pmatrix} \leq \begin{pmatrix} -1 \\ 1 \\ 3 \end{pmatrix}$$

has two quantifier changes. Figure 2 shows a visualization of the constraint polyhedron restricted to the unit cube. Since this example is rather small, we can guess a winning strategy for the existential player from the picture: 'Choose $x_1 \in [0, \frac{1}{2}]$,

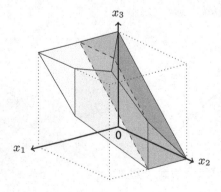

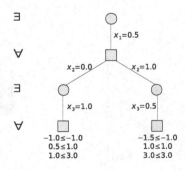

Fig. 2. Solution space of Example 2   **Fig. 3.** Winning strategy of Example 2

then choose x_3 appropriate to x_2, e.g., $x_3 = 1 - x_2$.' The highlighted solution space visualizes the set of all games with a definite winning outcome for the existential player. Figure 3 shows a winning-strategy for the existential player.

If there is more than one winning strategy for the existential player, it can be reasonable to search for a certain (the 'best') one. We can therefore modify the problem to include a linear objective function as shown in the following (where we note that transposes are suppressed when they are clear from the context to avoid excessive notation).

Definition 5 (QLPs with Objective Function). *Let $\mathcal{Q} \circ x \in [l, u] : Ax \leq b$ be given as in Definition 1 with the variable blocks being denoted by B_i. Let there also be a vector of objective coefficients $c \in \mathbb{Q}^n$. We call*

$$z = \min_{B_1}(c^1 x^1 + \max_{B_2}(c^2 x^2 + \min_{B_3}(c^3 x^3 + \max_{B_4}(\ldots \min_{B_m} c^m x^m)))))$$
$$\mathcal{Q} \circ x \in [l, u] : Ax \leq b \qquad \text{(QLP*)}$$

a QLP with objective function (for a minimizing existential player).

Note that the variable vectors $x^1, ..., x^i$ are fixed when a player minimizes or maximizes over variable block B_{i+1}. Consequently, it is a dynamic multistage decision process, similar as it is also known from multistage stochastic programming [3]. However, whereas in the latter an expected value is minimized, in our case we try to minimize the possible worst case (maximum loss) scenario that can result from the universal player's decisions. In the following we use the abbreviation $\min c^T x$ for the objective function and denote by

$$\min\{c^T x \mid \mathcal{Q} \circ x \in [l, u] : Ax \leq b\}$$

a quantified linear program with objective function.

Definition 6 (QLP Optimization Problems). *Given a QLP with objective function, the problem "Is it feasible? If yes, what is the best objective value of the existential player's winning policies?" is called* QLP Optimization Problem.

Note, that to find the existential player's optimal objective value, it is not sufficient to fix the universal players variables to their worst-case values regarding the objective function. For clarification, consider the following program:

$$\min\{-x_1 - 2x_2 \mid \forall x_1 \in [0, 1] \ \exists x_2 \in [0, 1] : x_1 + x_2 \leq 1\}$$

Judging from the objective function the universal player should fix $x_1 = 0$, but this results in a better objective value for the existential player than forcing the existential player to fix $x_2 = 0$ in the constraint system.

3   Nested Benders Decomposition (NBD)

The algorithm we extend in this paper was first proposed in [5] and uses decomposition techniques to solve an implicit reformulation of a QLP, which we call a *deterministic equivalent problem* (DEP). The concept is similar to the notion of deterministic equivalence as it is known from stochastic programming (cf. [3]) in the context of multistage stochastic linear programs (MSSLPs). Using the assumption of a finite time horizon and a discrete probability space, the resulting scenario tree is encoded into a DEP by replicating the LP for each possible scenario (possible path of events). Additionally, it is required that decisions must not depend on future events (*nonanticipativity*). The DEP of a QLP instance can be constructed in a similar way, however, instead of encoding the scenario tree of randomly arising scenarios, we encode the decision tree of the universal player, which results from the series of all possible upper and lower bound combinations of the universal variables as determined by the quantification sequence. Nodes at stage t are decision points where the existential player has to fix variables, e.g., by solving a linear program, with respect to all previous moves $(x^1, \ldots, x^{t-1})$. Arcs of the tree represent moves of the universal player when he fixes his variables to the corresponding lower and upper bounds. Figure 4(a) shows the universal player's decision tree for a QLP with quantification sequence $\exists x_1 \in [l_1, u_1] \, \forall x_2 \in [l_2, u_2] \, \exists x_3 \in [l_3, u_3] \, \forall x_4 \in [l_4, u_4] \, \exists x_5 \in [l_5, u_5]$. The tree is similar to the strategy of a QLP where the moves of the existential player have not been fixed.

Figure 4(b) shows the resulting DEP matrix structure using a compact variable formulation, which implicitly satisfies the nonanticipativity property because all nodes in the tree that share a common history also have the same set of decision variables up to that point. The resulting DEP grows exponentially with the number of universally quantified variables of the corresponding QLP, but the special block structure of the matrix can be exploited by the Nested Benders Decomposition (NBD) algorithm. The NBD algorithm is a recursive application of the well-known Benders Decomposition principle [2] and is widely used in the Stochastic Programming community to solve MSSLPs [3].

To illustrate how the Benders Decomposition algorithm works to solve QLPs, we consider w.l.o.g. the DEP that results from a QLP with quantification sequence $\exists x_1 \in [0, u_1] \forall x_2 \in [1, 1] \exists x_3 \in [0, u_3]$ and an objective function $\min c_1^T x_1 + c_3^T x_3$. Let the constraint system $A_1 x_1 + A_2 x_2 + A_3 x_3 \leq b$ contain the upper bound u_1

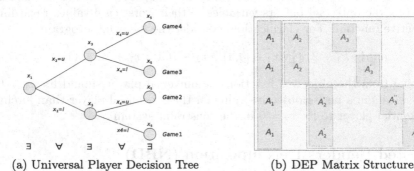

(a) Universal Player Decision Tree   (b) DEP Matrix Structure

Fig. 4. Deterministic equivalent linear program

of x_1 and u_3 of x_3. Since the $\forall$-variable $x_2 \in [1, 1]$ is a fixed variable, the DEP consists of a single game where the corresponding right hand side b' results from $b' = b - A_2\overline{x}_2$ with $\overline{x}_2 = 1$.

The resulting DEP looks as follows:

$$Z = \min c_1^T x_1 + c_3^T x_3$$
$$s.t. \qquad A_1 x_1 + A_3 x_3 \leq b'$$
$$x_1 \geq 0, x_3 \geq 0$$

Applying Benders Decomposition, the decision variables of the DEP are stage-wise partitioned and then decomposed into a *restricted master problem (RMP)* that contains the first-stage variable x_1, and one *subproblem (SP)* that contains the second-stage variable x_3. The corresponding *dual SP (DSP)* has the property that the solution space no longer depends on the value of x_1, regardless whether it is feasible or infeasible for the SP. The SP and its DSP can be written as follows:

$$SP(x_1) = \min c_3^T x_3 \qquad\qquad DSP(x_1) = \max \pi^T (b' - A_1 x_1)$$
$$s.t. \qquad A_3 x_3 \leq b' - A_1 x_1 \qquad s.t. \qquad A_3^T \pi \leq c_3$$
$$x_3 \geq 0 \qquad\qquad\qquad\qquad \pi \leq 0$$

For a non-optimal $\overline{x}_1$ obtained by solving the RMP, which can be empty at the beginning, the following two cases can happen. If the SP is feasible, the solution of the DSP is bounded and located at an extreme point of its solution space. If the SP is infeasible, the solution of the DSP is unbounded, which corresponds to an extreme ray of its solution space. Using this dual information, two different types of cutting planes - called *Benders cuts* - can be added to the RMP to cutoff the last $\overline{x}_1$ in the next solution of the RMP.

1. *feasibility cut*: $(\pi_r^j)(b' - A_1 x_1) \leq 0$, if the DSP$(\overline{x}_1)$ is unbounded, where π_r^j is the vector that corresponds to the extreme ray j.
2. *optimality cut*: $(\pi_p^i)(b' - A_1 x_1) \leq q$, if the DSP$(\overline{x}_1)$ is bounded, where π_p^i is the vector that corresponds to the extreme point i.

Since the DSP can only have finitely many extreme points and extreme rays, the RMP can be written as follows, where q is an auxiliary variable used to represent the objective function value of the SP:

$$RMP = \min c_1^T x_1 + q$$
$$s.t. \qquad (\pi_r^j)(d - A_1 x_1) \leq 0 \; \forall j \in J$$
$$(\pi_p^i)(d - A_1 x_1) \leq q \; \forall i \in I$$
$$x_1 \geq 0$$

This reformulation is equivalent to the initial DEP. However, there can be exponentially many extreme rays and extreme points and not all of them are needed to find the optimal solution. Therefore, the algorithm starts with I and J being empty, and computes cuts in an iterative process until an optimal solution is found or infeasibility is detected. In the latter case also the DEP and the corresponding QLP are infeasible. The optimal solution is found, if for a given candidate optimal solution (x_1^*, q^*), called proposal, also the SP(x_1^*) has an optimal solution with value $q(x_1^*)$ and the optimality condition $q(x_1^*) = q^*$ is satisfied. If this is the case, the algorithm stops. Otherwise a feasibility or optimality cut is added to the RMP, which is then re-solved again to obtain a new proposal. In each iteration where the SP is feasible, $c^T x_1^* + q^*$ yields a lower bound for the initial problem, while $c^T x_1^* + q(x_1^*)$ yields an upper bound. The difference between these bounds gets smaller, and if it becomes less than a predefined ϵ, the algorithm terminates.

If there are k universally quantified variables, then there are 2^k games and therefore 2^k subproblems are solved in each iteration, each yielding a cut that is added to the RMP. The *min-max* property of the objective function is achieved, because all optimality cuts that result from the subproblems restrict the same auxiliary variable q. For the computation of the upper bound, the maximum over all subproblems from the last iteration is used. For multistage QLPs resulting from a quantifier string $\exists\forall\exists\forall...\forall\exists$, Benders Decomposition can be recursively applied, which is known as Nested Benders Decomposition. Solving the DEP of a multistage QLP can be illustrated as solving a tree of linear programs that are attached to the nodes of the decision tree of the universal player. The tree is traversed forwards and backwards multiple times, with information being passed between adjoined nodes of the tree. A node at stage t passes proposals for the variables from the root up to stage t to its immediate descendants at stage $t+1$ and cuts to its immediate ancestor at stage $t-1$.

The algorithm has been implemented and tested in a detailed computational study with instances that were generated from existing LP and IP test sets [5].

4   Game-Tree Search and the $\alpha\beta$-Algorithm

The term *minimax tree* describes one of the most important data structures that allows computers to play two-person zero-sum games such as checkers, chess, and go. Nodes of the tree are decision points for the players and are therefore subdivided into min and max nodes. Nodes from different stages are connected with branches, leaf nodes are end positions of the game and can be evaluated as a win, loss, or draw using the rules of the game. Often, a specific score from the max player's point of view is computed with the help of a weighting function and assigned to a leaf to represent how good or bad the sequence of moves from the root to the leaf is. With a complete game-tree, it is possible to solve the game with the *MiniMax*-Algorithm, which fills the inner node values of the tree bottom-up starting with the evaluated values at the leafs. Nodes that belong to the max player get the maximum values of their successors, while nodes for the min player get the minimum. Figure 5(a) illustrates this behavior.

While the MiniMax-Algorithm must evaluate the entire game-tree to compute the root value, the $\alpha\beta$-*Algorithm* [8,11] prunes away branches that cannot influence the final result. Therefore, it maintains two values, α and β, which represent the minimum score that the max player is sure to gain at least until

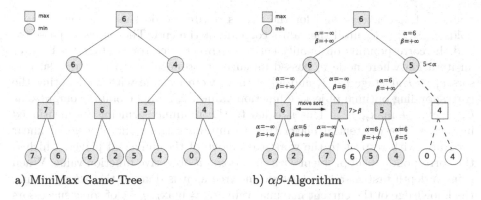

a) MiniMax Game-Tree   b) $\alpha\beta$-Algorithm

Fig. 5. MiniMax game-tree and $\alpha\beta$-algorithm

that point in the tree, and the maximum score of the min player respectively. If the evaluation of a position where the min player has to move becomes less than α, the move need not to be further explored, since a better move has already been found. The same holds, if at a position where the max player has to choose its move, the evaluation provides a value that is greater than β. Figure 5(b) illustrates this behavior, the dashed subtrees were not visited. The left one due to a β-cutoff, the subtree on the right hand due to an α-cutoff.

While the order in which the nodes of the tree are evaluated does not matter for the MiniMax-Algorithm, it is essential for the performance of the $\alpha\beta$-Algorithm.

The *best moves* need to be evaluated first in order to find strong α and β values as soon as possible. Figure 5(b) illustrates this, without swapping the subtrees under the first successor of the root on the left side, the β-cutoff would not have occurred. If the best moves are searched first, the runtime of the $\alpha\beta$-Algorithm is only $O(\sqrt{b^d})$ where d is the depth of the tree and b is the branching factor, i.e., the number of possible moves at each node. The MiniMax-Algorithm has a runtime of $O(b^d)$.

5   The $\alpha\beta$-Nested Benders Decomposition ($\alpha\beta$-NBD)

In this Section we describe how the $\alpha\beta$-technique in combination with move-ordering can be integrated into the Nested Benders Decomposition (NBD) algorithm. Let us recall that solving a multistage QLP with the NBD algorithm can be illustrated as solving a tree of linear programs that are attached to the nodes of the universal player's decision tree. The tree is therefore traversed multiple times and information in the form of proposals and cuts are passed between nodes of the tree. If a node v_i at stage $t \in \{0, \ldots, T\}$ receives a new proposal $\overline{x}^{t-1}$ from its direct successor at stage $t-1$, the subtree rooted at node v_i is solved to optimality, or until the nodal linear program attached to v_i becomes infeasible. After feasibility of the subtree is established, the upper and lower bounds of v_i converge and for the node's optimal objective function value z_i holds $L_i \leq z_i \leq U_i$ until the values coincide at the end. Then node v_i passes $\overline{z}_i$ and the corresponding optimality cut to its direct ancestor at stage $t - 1$. After an iteration where node v_i passed its current proposal $\overline{x}^t$ to its direct successors $v_j \in J$ at stage $t + 1$, and all of them were feasible with $\overline{z}_j$ denoting the corresponding optimal objective function values, v_i's upper bound computes as $U_i = c^T \overline{x}^t + \max_{j \in J} \overline{z}_j$. This is equal to the minimax principle as mentioned in Sect. 4. The existential player tries to minimize the value of a nodal linear program, with respect to the worst-case move of the universal player, which is the corresponding subproblem with the *maximal* objective function value. When using a depth first-search to traverse the tree as it is done in the $\alpha\beta$-algorithm, the knowledge of the current maximal value $\alpha_i = \max_{k \in K} \overline{z}_k$ of some successors $v_k \in K \subset J$ of node v_i can be used in a similar manner as α in the $\alpha\beta$-algorithm. In the current iteration, it denotes the minimum value, the maximizing player (the universal one) will at least obtain at node v_i. When α_i is passed to the remaining nodes from the set of successors $J \backslash K$ each node v_l from this set can stop computing its exact optimal objective function value after it determines feasibility with respect to the current proposal $\overline{x}^t$, and detects that its current upper bound U_l is less than or equal to α_i. We can also integrate a value analogously to β that depicts the maximum value the minimizing player will gain for sure at a specific node v_i at stage t. In terms of the NBD algorithm this is the upper bound U_i from the previous iteration. In the next iteration, this value can be passed to all successors $v_j \in J$ at stage $t + 1$ together with the new proposal $\overline{x}^t$. If a successor v_j determines feasibility with respect to the current proposal $\overline{x}^t$ and detects that its lower bound $L_i = c^T \overline{x}^{t+1} + \overline{q}^{t+1}$ is greater than this value, it can stop computing its exact optimal objective function value because a better solution has already been found in the previous iteration. In the following, we will therefore also use the abbreviations α and β for these values.

As in the case of the $\alpha\beta$-algorithm, the order in which the nodes of the tree are solved is an important issue in the $\alpha\beta$-NBD algorithm. Whereas the $\alpha\beta$-algorithm uses a heuristic evaluation function, our algorithm organizes the order in which nodes are visited based on information from previous iterations. To obtain strong bounds as soon as possible, the successor of a node v_i that provided the worst-case sub solution in the previous iteration, is visited first in the next iteration, speculating that it will again provide a strong α-bound. Also, the other successors are arranged in descending order by their solution values from the last iteration. However, many other sorting criterions are possible.

The algorithmic framework has been implemented in C++ using the LP Solver CPLEX 12.4 to solve nodal linear programs.

6 Computational Results

In the following we present the results of our experimental evaluation. All tests were run on a quad-core processor AMD Phenom II X4 945 with 8GB RAM. For our tests we took LP instances with a maximum number of 500 variables

Table 1. Computational results

∀-Vars	∀-Blocks	CPLEX Time (s)	NBD Time (s)	Subproblems	NBD ($\alpha\beta$ + move-ordering) Time (s)	Subproblems
10	1	572.84	36.79	123844	21.72	82832
10	2	225.22	69.48	300410	32.51	137550
10	5	109.65	101.20	567935	53.96	334402
15	1	>172800.00	1375.60	5923805	1010.70	4117858
15	2	>172800.00	1755.77	7592959	1101.67	4578949
15	5	130934.78	2281.55	13021052	1113.95	6724840
18	1	>172800.00	10044.65	38954373	7047.60	24542092
18	2	>172800.00	15174.01	75414983	10059.71	48808515
18	5	>172800.00	27007.41	123214174	14313.93	66535654

and constraints and generated QLPs with 10, 15, and 18 universally quantified variables. For each new universally quantified variable $x_i \in [0,1]$, we randomly added matrix coefficients from the interval $[-1,1]$ with a density of 25 %. We furthermore varied the number of ∀-quantifier blocks to 1, 2, and 5 and distributed them equally in the QLP. This results in twostage and multistage QLP instances and nine different test sets. A similar test set was used in [5].

Table 1 shows the summed up results solving each of the test sets with the standard NBD algorithm, the $\alpha\beta$-Nbd algorithm, and when solving the corresponding DEPs with CPLEX. Column 1 contains the number of universally quantified variables followed by the number of blocks of universally quantified variables in column 2. Column 3 contains the solution times when the corresponding DEPs are solved with CPLEX running with standard settings and its preprocessor enabled. Columns 4 and 5 show the solution times and the number of LPs that were solved using the standard NBD algorithm. Columns 6 and 7 show the same numbers when $\alpha\beta$-cuts and the move sort heuristic are used.

The results show that the even the standard NBD algorithm implementation is clearly faster than solving the DEP in most cases, especially with an increasing number of universally quantified variables. This is due to the exponential growth of the DEP with an increasing number of universally quantified variables in the corresponding QLP. When we additionally use the $\alpha\beta$-heuristic values and move sort, we observe notable time savings up to about 50 % compared to the standard implementation as we can, e.g., see in the last row of Table 1. The extended algorithm was able to halve the number of subproblems that had to be solved from $123,214,174$ to $66,535,654$, resulting in a reduction of the solution time from 450 min to 238 min, a difference of 47 %. The effect becomes stronger with an increasing number of stages but even in the twostage case, move-ordering alone leads to a performance gain of 25 %–40 %. These results show the high potential of combining techniques from game-tree search with the Nested Benders Decomposition approach and motivate a further research in this direction.

7   Outlook: Application to Other Games

Apart from the fact that QLPs taken by themselves can be interpreted as two-person zero-sum games, they provide an adequate modeling tool for many purposes. In [6] a QIP model of the two-person game Gomoku was proposed. We adopted this model to a rather similar PSPACE-complete game: Connect6. Here two players playing on a Go board try to achieve a connected row of six stones. At the beginning, black places one stone, then white and black take turns placing two stones. The player, who has the first connected row of six stones, wins.

While the former model was never practically solved, we use only a logarithmic number of universally quantified variables in our current model to reduce the computational burden. This is necessary because the effort to solve a QIP with both DEP and $\alpha\beta$-NBD grows exponentially with the number of universally quantified variables. To model a set of n binary variables x_i where exactly one equals 1 while all others are 0 (similar to a so called SOS1-constraint in mathematical optimization) with a logarithmic amount of binary variables y_i, one can use the following transformation between binary and unary encoding[1]:

$$\sum_{i=1}^{\log_2(n)} 2^{i-1} \cdot y_i = \sum_{i=1}^{n} i \cdot x_i - 1$$

$$\sum_{i=1}^{n} x_i = 1$$

Still the DEP becomes too large to solve directly. Thus we used a variant of the proposed $\alpha\beta$-NBD algorithm to decide whether black can win in n moves starting from an arbitrary situation. However, as the model contains binary variables, an additional type of cut, called Combinatorial Benders Cut [17], had to be added to the algorithm. Given a node v_i at depth t a proposal $\overline{x}^{t-1}$ that turns out to be invalid, at least the corresponding part of the solution space of the master problem is cut off by the following cut:

$$\sum_{j\in\Lambda:\overline{x}_j^{t-1}=0} \overline{x}_j^{t-1} + \sum_{j\in\Lambda:\overline{x}_j^{t-1}=1} (1 - \overline{x}_j^{t-1}) \geq 1$$

Here Λ depicts the set of existentially quantified variables from stage 0 to stage $t-1$. Our plan is to improve this cut by methods of conflict analysis in MIPs.

In our preliminary tests, we could solve instances of this problem for $n \leq 10$ on a board of size 8×8.

8   Summary

In the course of this paper we considered QLPs with objective functions and showed how their hybrid nature of being a two-person zero-sum game on the

---

[1] Heed that this transformation is only valid if n is a power of 2. It can be easily adopted to the general case.

one side, and being a convex multistage decision problem on the other side, can be used to combine linear programming techniques with solution techniques from game-tree search. We therefore extended the Nested Benders Decomposition algorithm by the $\alpha\beta$-algorithm in combination with its heuristic move-ordering, as used in evaluating minimax trees. We showed the applicability in an experimental evaluation, where we solved QLPs that were generated from the well-known Netlib test set. The results showed a speedup of up to 50 % compared to the standard Nested Benders Decomposition implementation without techniques from game-tree search.

Acknowledgements. This research is partially supported by German Research Foundation (DFG) funded SFB 805 and by the DFG project LO 1396/2-1.

References

1. Allis, L.V., van der Meulen, M., van den Herik, H.J.: Proof-number search. Artif. Intell. **66**(1), 91–124 (1994)
2. Benders, J.F.: Partitioning procedures for solving mixed-variables programming problems. Numer. Math. **4**(1), 238–252 (1962)
3. Birge, J.R., Louveaux, F.: Introduction to Stochastic Programming. Springer Series in Operations Research and Financial Engineering. Springer, New York (1997)
4. Donninger, C., Lorenz, U.: The hydra project. Xcell J. (53), 94–97 (2005)
5. Ederer, T., Lorenz, U., Martin, A., Wolf, J.: Quantified linear programs: a computational study. In: Demetrescu, C., Halldórsson, M.M. (eds.) ESA 2011. LNCS, vol. 6942, pp. 203–214. Springer, Heidelberg (2011)
6. Ederer, T., Lorenz, U., Opfer, T., Wolf, J.: Modeling games with the help of quantified integer linear programs. In: van den Herik, H.J., Plaat, A. (eds.) ACG 2011. LNCS, vol. 7168, pp. 270–281. Springer, Heidelberg (2012)
7. Hsu, F.H.: Ibm's deep blue chess grandmaster chips. IEEE Micro **18**(2), 70–80 (1999)
8. Knuth, D.E., Moore, R.W.: An analysis of alpha-beta pruning. Artif. Intell. **6**(4), 293–326 (1975)
9. Lorenz, U., Martin, A., Wolf, J.: Polyhedral and algorithmic properties of quantified linear programs. In: de Berg, M., Meyer, U. (eds.) ESA 2010, Part I. LNCS, vol. 6346, pp. 512–523. Springer, Heidelberg (2010)
10. Papadimitriou, C.H.: Games against nature. J. Comput. Syst. Sci. **31**, 288–301 (1985)
11. Pijls, W., de Bruin, A.: Game tree algorithms and solution trees. Theor. Comput. Sci. **252**(1–2), 197–215 (2001)
12. Plaat, A., Schaeffer, J., Pijls, W., De Bruin, A.: Best-first fixed-depth game-tree search in practice. In: Proceedings of the 14th International Joint Conference on Artificial Intelligence, San Francisco, CA, USA, vol. 1, pp. 273–279. Morgan Kaufmann Publishers Inc. (1995)
13. Subramani, K.: Analyzing selected quantified integer programs. In: Basin, D., Rusinowitch, M. (eds.) IJCAR 2004. LNCS (LNAI), vol. 3097, pp. 342–356. Springer, Heidelberg (2004)
14. Subramani, K.: On a decision procedure for quantified linear programs. Ann. Math. Artif. Intell. **51**(1), 55–77 (2007)

15. van den Herik, H.J., Nunn, J., Levy, D.: Adams outclassed by hydra. ICGA J. **28**(2), 107–110 (2005)
16. van den Herik, H.J., Uiterwijk, J.W.H.M., van Rijswijk, J.: Games solved: now and in the future. Artif. Intell. **134**, 277–312 (2002)
17. Vanderbeck, F., Wolsey, L.: Reformulation and decomposition of integer programs. CORE Discussion Papers 2009016, Université catholique de Louvain, Center for Operations Research and Econometrics (CORE) (2009)
18. Winands, M.H.M., Uiterwijk, J.W.H.M., van den Herik, H.J.: PDS-PN: a new proof-number search algorithm. In: Schaeffer, J., Müller, M., Björnsson, Y. (eds.) CG 2002. LNCS, vol. 2883, pp. 61–74. Springer, Heidelberg (2003)

Improving Best-Reply Search

Markus Esser, Michael Gras, Mark H.M. Winands$^{(\boxtimes)}$,
Maarten P.D. Schadd, and Marc Lanctot

Games and AI Group, Department of Knowledge Engineering,
Maastricht University, Maastricht, The Netherlands
markus.esser1@rwth-aachen.de, michael.gras@gmx.net, m.schadd@gmail.com,
{m.winands,marc.lanctot}@maastrichtuniversity.nl

Abstract. Best-Reply Search (BRS) is a new search technique for game-tree search in multi-player games. In BRS, the exponentially many possibilities that can be considered by opponent players is flattened so that only a single move, the best one among all opponents, is chosen. BRS has been shown to outperform the classic search techniques in several domains. However, BRS may consider invalid game states. In this paper, we improve the BRS search technique such that it preserves the proper turn order during the search and does not lead to invalid states. The new technique, BRS$^+$, uses the move ordering to select moves at opponent nodes that are not searched. Empirically, we show that BRS$^+$ significantly improves the performance of BRS in Four-Player Chess, leading to winning 8.3 %–11.1 % more games against the classic techniques max^n and Paranoid, respectively. When BRS$^+$ plays against max^n, Paranoid, and BRS at once, it wins the most games as well.

1 Introduction

Research in the field of artificial intelligence has enjoyed immense success in the area of two-player zero-sum games of perfect information. Well-known examples of such progress include IBM's DEEP BLUE vs. Kasparov [1], self-play learning to reach master level in Backgammon [11], and solving the game of Checkers [8].

Interest in abstract, deterministic multi-player games (>2 players) has grown [5,7,10], but the amount of research in this area remains relatively small in comparison to that in the two-player setting. This is partly due to the fact that there are no worst-case equilibrium guarantees, but also to the added complexity introduced by more than two players.

In this paper, we propose a multi-player search technique, called BRS$^+$, which improves on a previous search technique called Best-Reply Search (BRS) [7]. In BRS, the root (search) player enumerates each of his[1] moves, but search at opponents' nodes is restricted: only one of the opponents is allowed to act, the others must pass sequentially and the best single decision among all opponents is chosen to be played against the root player. As a result, the turn sequence

---

[1] For brevity, we use 'he' and 'his' whenever 'he or she' and 'his or her' are meant.

H.J. van den Herik et al. (Eds.): CG 2013, LNCS 8427, pp. 125–137, 2014.
DOI: 10.1007/978-3-319-09165-5_11, © Springer International Publishing Switzerland 2014

is flattened so that the decisions strictly alternate between the root player and an opponent player. Collapsing the opponents' decisions in this way can lead to invalid game states due to inconsistent turn order. For example, passing is not allowed in many classic games studied by AI researchers. Also, this modified search allows the root player to make more moves along the search path as all of the opponents are combined. Despite these problems, the computational load of the search is reduced and BRS has been shown to work well in practice in several different multi-player games [3, 6, 7].

The main contribution of this paper is a reformulation of BRS that (1) ensures only valid states are reached and that (2) the proper turn sequence is preserved, while still retaining the benefit of reduced computational complexity. This variant, called BRS$^+$, selects moves using move orderings rather than passing. We show that BRS$^+$ performs significantly better than BRS and the classic multi-player search techniques max^n [5] and Paranoid [10] in Four-Player Chess [4, 12].

The organization of the paper is as follows. First, we introduce the game of Four-Player Chess in Sect. 2. Then we formalize our problem and describe previous work in Sect. 3. We introduce BRS$^+$ and analyze its complexity in Sect. 4. We show results from a variety of experiments in Sect. 5. Finally, we conclude the paper and discuss potential future work in Sect. 6.

2   Four-Player Chess

Four-Player Chess is an extension of the classic game to four players. Its initial position is shown in Fig. 1a. The rules we use here are adapted from the Paderborn rule set [4, 12]. Players strictly alternate turns clockwise, starting with White (1) to move, then Red (2), then Black (3), then Blue (4). Most rules from the two-player game remain unchanged, and the main differences are as follows.

- Pawns are allowed to *bend-off* into a new direction at the diagonals such that the distance to promotion is kept the same as normal. This allows promotions on all sides. Once a pawn has bent-off, its direction can no longer change.
- En-passant can sometimes be combined with a capture move resulting in two captured pieces. For example, Player 2 moves a pawn that enables an en-passant move for Player 4 and Player 3 moves a piece to the en-passant capture square. In Fig. 1b, Player 2 moves the pawn two spaces, Player 3 moves the rook, enabling an *en-passant double capture* for Player 4.
- Players can be eliminated in two ways: either they are check-mated at the beginning of the turn, or their king is hanged. The king hanging occurs for a player when that player cannot respond to his king being put in check due to turn order. In Fig. 1c, if Player 1 moves his bishop and Player 2 is next to play, Player 3 is eliminated immediately. Once a player is eliminated, all of his pieces are removed from the board.
- The winner is the last player standing.

The game is made to be finite by forcing the usual draw-by-repetition and 50-move rule, each being unchanged from the standard 2-player version.

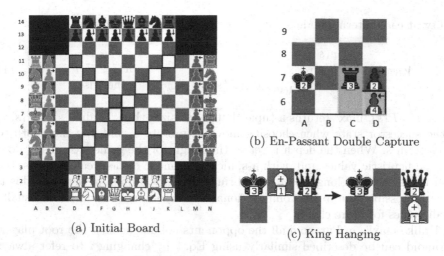

(a) Initial Board

(b) En-Passant Double Capture

(c) King Hanging

Fig. 1. Four-Player Chess (Color figure online)

3 Multi-player Search

This section discusses search techniques for deterministic multi-player games. First, the classic search techniques max^n and Paranoid are described in Subsect. 3.1. Next, Best-Reply Search (BRS) is introduced in Subsect. 3.2.

3.1 Max^n and Paranoid Search

A finite game of perfect information can be described by a tuple $(\mathcal{N}, \mathcal{S}, \mathcal{Z}, \mathcal{A}, \mathcal{T}, P, u_i, h_i, s_0)$. The player set $\mathcal{N} = \{1, \dots, n\}$ contains player labels and by convention a player is denoted $i \in \mathcal{N}$. The state space $\mathcal{S}$ is a finite, non-empty set of states, with $\mathcal{Z} \subseteq \mathcal{S}$ denoting the finite, non-empty set of terminal states. The move set $\mathcal{A}$ is a finite and non-empty set. The utility functions $u_i : \mathcal{Z} \mapsto [v_{\min}, v_{\max}] \subseteq \mathbb{R}$ gives the utility of Player i, with v_{min} and $v_{\max}$ denoting the minimum and maximum possible utility, respectively. The heuristic evaluation functions $h_i : \mathcal{S} \mapsto [v_{\min}, v_{\max}]$ return a heuristic value of a state. In this paper, we assume constant-sum games: $\forall z \in \mathcal{Z}, \sum_{i \in \mathcal{N}} u_i(z) = k$, for some constant k. The player index function $P : \mathcal{S} \to \mathcal{N}$ returns the player to act in a given non-terminal state s, or a null value when its argument is a terminal state. The transition function $\mathcal{T} : \mathcal{S} \times \mathcal{A} \mapsto \mathcal{S}$ describes the successor state given a state s and a move chosen from the available moves $\mathcal{A}(s) \subseteq \mathcal{A}$. We will also refer to the null or pass move as $\emptyset$. The game starts in an initial state s_0 and with player $P(s_0)$ to act. Finally, the tuple $u(s) = (u_1(s), \cdots, u_n(s))$ and $h(s)$ are defined in a similar way.

There are two classic techniques used for game-tree search in multi-player games: max^n [5], and Paranoid [10]. Given a root state s, max^n searches to a fixed depth, selecting the move that maximizes the acting player's individual utility at each internal node $s \in \mathcal{S} \setminus \mathcal{Z}$, defined for move a at state s by:

$$V_d(s,a) = \begin{cases} u(s') & \text{if } s' \in \mathcal{Z} \\ h(s') & \text{if } s' \notin \mathcal{Z}, d = 0 \\ \max_{a' \in \mathcal{A}(s')}^i V_{d-1}(\mathcal{T}(s',a'),a') & \text{otherwise,} \end{cases} \quad (1)$$

where $i = P(s)$, $\max^i$ returns a tuple that maximizes the ith value, $s' = \mathcal{T}(s,a)$ is the successor state when choosing move a in state s, and d is the depth to reach from s. When the depth is set to the height of the game tree, $\max^n$ does not use heuristic values and, with a sufficient number of tie-breaking rules, the final move choice belongs to an equilibrium strategy profile. In practice, this is often impossible due to the computational time requirement and hence usually small values for d are chosen.

Unlike $\max^n$, in Paranoid all the opponents collude against the root player. Paranoid can be described similarly using Eq. 1 by changing i to refer always to the root player, and adding $\min_{a' \in \mathcal{A}(s')}^i V_{d-1}(s',a')$ if $s' \notin \mathcal{Z}, d > 0, P(s) \neq i$. In practice, consecutive opponent nodes can be thought of as one big opponent node, where each meta-move corresponds to a sequence of opponent moves. Hence, the game effectively becomes a two-player game, and $\alpha\beta$-style pruning can be applied.

Assuming a uniform branching factor b and depth d, the worst-case running time required for both classic search techniques is $O(b^d)$ since all nodes may require visiting. In the best case, Paranoid takes time $O(b^{d(n-1)/n})$ in an n-player game [10]. In practice, Paranoid tends to outperform $\max^n$ since it searches more deeply due to taking advantage of more pruning opportunities.

There are variants and hybrids of these techniques. For example, ProbMax^n incorporates beliefs from opponent models into the choice of decision made by $\max^n$. The Comixer technique models potential coalitions that can be formed from the current state of the game and chooses a move based on a mixed recommendation from each coalition [4]. MP-Mix decides on which search technique to invoke depending on how far the leading player is perceived to be ahead from the second-placed player [13].

3.2   Best-Reply Search

In Best-Reply Search (BRS), instead of enumerating every move at each of the opponents' nodes independently in sequence, the opponents' nodes and moves are merged into one opponent decision node [7]. This is similar to Paranoid with one key difference: the moves at the opponent nodes belong to *one* of the $(n-1)$ opponents. The particular move chosen at the opponent decision node is one that minimizes the root player's utility. As a result, the nodes visited strictly alternate between the root player's (max) nodes and the opponent decision (min) nodes. In other words, only a single opponent chooses a move and the other opponents pass. The tree transformation process is depicted in Fig. 2. Referring to Fig. 2a, in general $\{g,h\} \neq \{i,j\}$ and similarly $\{k,l\} \neq \{m,n\}$.

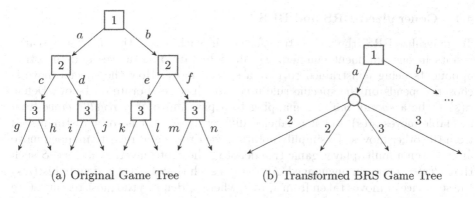

(a) Original Game Tree   (b) Transformed BRS Game Tree

Fig. 2. An example of how BRS transforms the multi-player search tree. In the BRS tree, the edges labeled 2 assume a pass move by Player 3 and vice versa.

Assume now Player 1 chooses a and Player 2 ignores his legal moves and passes (plays $\emptyset$), there are no longer two distinct nodes that belong to Player 3. The game has then reached a state $s_{a,\emptyset} = \mathcal{T}(\mathcal{T}(s,a),\emptyset)$. As a result, there is a single move set $\mathcal{A}(s_{a,\emptyset})$ available to Player 3. Define $s_{a,c}$ and $s_{a,d}$ similarly. To simplify the analysis, we assume uniform game trees, so in this example $|\mathcal{A}(s_{a,\emptyset})| = |\mathcal{A}(s_{a,c})| = |\mathcal{A}(s_{a,d})| = b$. Clearly, there will be duplicate moves at the opponent nodes in the BRS tree in Fig. 2b, which need not be searched more than once. In fact, the opponent who is allowed to search will have b moves. There are $(n-1)$ opponents, therefore there will be $b(n-1)$ unique opponent choices to consider at opponent nodes rather than b^{n-1} in the case of Paranoid and $\max^n$. As a result, BRS requires less time for a depth d search than Paranoid and $\max^n$. However, this benefit comes at the cost of an approximation error since collapsing the opponents' decisions in this way can lead to invalid game states due to an inconsistent turn order, as passing is not allowed in many abstract games. Also, this modified search allows the root player to make more moves as all of the opponents are combined. Despite these problems, the computational load of the search is reduced and BRS has been shown to work well in practice in several different multi-player games such as Billabong [3], Blokus [6], Chinese Checkers [6,7], and Focus [6,7].

4   Generalized Best-Reply Search

In this section we present a reformulation of Best-Reply Search (BRS) by suggesting a different transformation of the game tree. The resulting BRS⁺ tree is an augmented sub-graph of the original game tree. Then, applying the usual Paranoid-style search over in this new tree results in BRS⁺ (Subsect. 4.1). Next, a complexity analysis is given in Subsect. 4.2.

4.1   Generalized BRS and BRS$^+$

To generalize BRS, the tree is transformed in such a way that a regular search results in one opponent enumerating all of his available moves and the other opponents being constrained to play a special move. How the special move is chosen depends on the specific rule used, which we elaborate on below. Define $\phi(s)$ to be a *special move*: a mapping to a pass move or a *regular move* from available moves $\mathcal{A}(s)$, but it is labeled differently since it is to be distinguished from the other moves. To simplify notation, we drop s and refer to mapped move as ϕ. Given a multi-player game tree denote i the root player. States $s \in \mathcal{S}$ such that $P(s) = i$ remain unchanged. At states s such that $P(s) \neq i$, denote hist(i, s) the sequence of moves taken from s_i to s, where s_i denotes the most recent[2] state such that $P(s_i) = i$. If hist(i, s) contains exactly two regular moves, then all the regular moves allowed have been taken along hist(i, s), so set $\mathcal{A}'(s) = \{\phi\}$. Otherwise, set $\mathcal{A}'(s) = \mathcal{A}(s) \cup \{\phi\}$ unless $\mathcal{T}(s, a) = i$ in which case the node remains unchanged[3]. This last condition is required to ensure that the number of regular moves along hist$(i, \mathcal{T}(s, a))$ is always equal to 2. This transformation with $i = 1$ applied to the game tree in Fig. 2a is shown in Fig. 3.

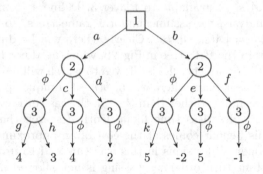

Fig. 3. An example of how the generalized BRS models the multi-player search tree, with payoffs belonging to the root player.

This construction with the special moves generalizes BRS. For example, if $\phi = \emptyset$ the generalized BRS tree is equivalent to the one in Fig. 2b. We focus on mapping special moves to move ordering moves (MOMs). In this case, illegal states cannot be reached since ϕ maps to a move in $\mathcal{A}(s)$, and we refer to the algorithm as BRS$^+$. The trees are not actually transformed. The move sets are manipulated during the recursive search, as presented in Algorithm 1. The critical modification in the tree search is to ensure that for every sequence of moves starting and ending at the root player, exactly one regular move and $(n - 2)$ special moves are taken by the opponents. This is achieved by counting

---

[2] By "most recent" we mean the path with the shortest such sequence of moves.

[3] We assume, without loss of generality, that the game has a strictly alternating turn order, so $P(\mathcal{T}(s, a))$ will be the same $\forall a \in \mathcal{A}(s)$.

```
1  GBRS(node s, depth d, count m, root player i)
2      if s ∈ Z then return u_i(s)
3      ;
4      else if d = 0 then return h_i(s)
5      ;
6      ;
7      else
8          A'(s) ← A(s)
9          Let U ← ∅ be the set of child values, and j be the player following P(s)
10         if P(s) = i then m ← 0
11         ;
12         else if P(s) ≠ i and m = 2 then A'(s) ← {φ}
13         ;
14         ;
15         else if P(s) ≠ i and j ≠ i then A'(s) ← A(s) ∪ {φ}
16         ;
17         ;
18         for a ∈ A'(s) do
19             s' ← T(s, a); m' ← m if a = φ else m + 1
20             u' ← GBRS(s', d − 1, m', i)
21             U ← U ∪ {u'}
22         return max(U) if P(s) = i else min(U)
```

Algorithm 1. Generalized Best-Reply Search

the regular moves between successive turns of the root player i using a parameter m in Algorithm 1.

Two different approaches can be taken for selecting a special move by move ordering. One can order the moves in a way that attacks the root player i, or order moves in a way that looks most promising to $P(s)$. The former *paranoid move ordering* can be too pessimistic, preferring to capture the root player's pawn instead of another opponent's queen. The latter max^n *move ordering* can prefer to capture another player's queen over the root player's bishop even though the root player is in a much better position to win the game. We analyze the effect of these specific move-ordering strategies in Sect. 5.

Similarly to Paranoid, only the root player's payoff is being considered, so standard $\alpha\beta$ pruning can be applied in BRS and BRS$^+$. From Fig. 3, after the left subtree is traversed the value of taking move a is assessed to be 2. When the opponents are minimizing in the right subtree and the -2 is found, the remaining branches can be pruned since the root player will never choose move b since move a already has a higher value. Finally, when the special moves are mapped to MOMs, transpositions will occur, which can be subsequently be pruned [3].

4.2   Complexity Analysis

To show the number of nodes expanded in BRS$^+$, we use a similar argument to the one for analyzing Paranoid [10]. Recall that, as is common in this setting, uniform trees are assumed and hence the branching factor, b, is a constant.

We start with the worst-case analysis. To analyze the complexity, we return to the original tree transformation depicted in Fig. 2b (specifically, *not* the tree depicted in Fig. 3.) The same argument used in Subsect. 3.2 to merge the state and move sets applies when the opponent moves are mapped using ϕ. We first analyze the depth reached in this transformed tree and then relate it to the true depth reached. Assume depth D is reached in this tree, then $D/2$ levels of max nodes are expanded and $D/2$ levels of min nodes are expanded. For simplicity, we assume D is even. How does D relate to the true depth d? At a max node in the tree, one ply represents a decrease in true depth of 1. At a min node, one ply represents a decrease in true depth of $(n-1)$. Therefore $d = D/2 + (n-1)D/2 \Rightarrow D = 2d/n$. At max nodes, the branching factor is b. At min nodes, the branching factor is $b(n-1)$. Therefore, the number of nodes expanded is $f_{worst}(b, n, D) = b \cdot b(n-1) \cdot b \cdot b(n-1) \cdots b(n-1)$, with $\frac{D}{2}$ occurrences of b and $\frac{D}{2}$ occurrences of $b(n-1)$. Formally, $f_{worst}(b, n, D) =$

$$b^{\frac{D}{2}} \cdot (b(n-1))^{\frac{D}{2}} = b^{\frac{D}{2}} \cdot b^{\frac{D}{2}} \cdot (n-1)^{\frac{D}{2}} = b^D \cdot (n-1)^{\frac{D}{2}} = b^{\frac{2d}{n}} \cdot (n-1)^{\frac{d}{n}},$$

so the time taken for a depth d search is $O(b^{\frac{2d}{n}}(n-1)^{\frac{d}{n}})$. Therefore, since $n > 2$, BRS$^+$ expands asymptotically fewer nodes as $d \to \infty$ than max^n and Paranoid in the worst case when $b > n - 1$, which is true in most large games.

The best-case analysis follows the original BRS best-case analysis [7]. There are two extremes. Assume the recursive search finds the lowest payoff value (a loss) at every leaf visited. Every min node searches a single move in this case because the rest of the moves can be pruned due to a loss being the worst possible payoff. Max nodes still require enumerating each move in the hopes of finding something better than a loss. Using the same logic as above, this requires $b^{\frac{D}{2}} = b^{\frac{d}{n}}$ expansions. The other extreme is that every leaf visited is a win. In this case the roles are reversed and every move at max nodes after the first can be pruned but all the moves at min nodes need to be enumerated. This requires $(b(n-1))^{\frac{D}{2}} = b^{\frac{d}{n}} \cdot (n-1)^{\frac{d}{n}}$. The search technique must specify a strategy for all players, so if max has a win then min must have a loss, and vice versa. Therefore, the best case is the worst of these extremes, and hence $O(b^{\frac{d}{n}}(n-1)^{\frac{d}{n}})$ node expansions are required. As a result, compared to the worst case complexity, the exponent of b is halved.

While these results are encouraging, we make two key observations. First, the overhead involved with computing the move-ordering move may not be negligible. Secondly, the performance critically depends on the quality of the move ordering. If the move suggested is worse than passing, BRS$^+$ could perform worse than its predecessor, BRS. We investigate these points further in the following section.

5   Experimental Results

When directly comparing two multi-player search techniques, there are $2^4 - 2 = 14$ possible seating arrangements (four of the same type of player is not allowed).

When comparing four techniques there are $4! = 24$ possible seating arrangements and 36 when exactly three techniques are present. To ensure fairness, all of the experiment results are performed in batches where each batch consists of every possible seating arrangement for the number of players being compared. Therefore, the number of games per experiment is always a multiple of the number of seating arrangements for the specified number of players.

The Four-Player Chess engine is written in Java, and the experiments were run on machines equipped with AMD Opteron 2.4 GHz processors. The evaluation function and search enhancements are described in [2]. All the times listed in this section are in milliseconds, and intervals represent 95 % confidence intervals.

5.1 Baseline Experiments

As a first experiment we ran 1440 games of the base search techniques max^n, Paranoid, and BRS. The results are listed in Table 1. As we can see from the results, Paranoid and BRS outperform max^n considerably, with Paranoid performing slightly better than BRS.

These results were unexpected, so we ran another experiment to compare each individual search technique directly. BRS beat max^n in 56.0 % ± 4.2 % of the games, BRS won against Paranoid in 58.9 % ± 4.1 % of the games, and Paranoid beat max^n in 59.0 % ± 2.9 % of the games [2]. These results suggest that while BRS is able to beat Paranoid, Paranoid does a slightly better job of exploiting max^n than BRS does, so when all three are present Paranoid has a slight edge over BRS.

Table 1. Performance results of BRS vs. max^n vs. Paranoid experiments.

Time (ms)	Games	BRS	max^n	Paranoid
1000	1440	38.4 % ± 2.6 %	19.6 % ± 2.1 %	42.1 % ± 2.6 %

Table 2. Direct comparison using different static move orderings for BRS$^+$.

Time (ms)	Games	BRS$^+$ (max^n MO)	BRS$^+$ (paranoid MO)
1000	1120	42.3 % ± 2.9 %	57.7 % ± 2.9 %

5.2 Move Ordering Experiments in BRS$^+$

In the next series of experiments we compare the two different approaches for selecting the special moves in BRS$^+$: paranoid vs. max^n move ordering. Recall the differences from Sect. 4. The paranoid move ordering prefers immediate captures of the root player's pieces while max^n move ordering considers captures of all opponents' pieces. The results of the two compared directly, as mentioned above, are shown in Table 2. As we see, the Paranoid move ordering performs better. This is somewhat expected as BRS^+ more closely resembles Paranoid than max^n. We will test the performance of each type of static move ordering against the other search techniques in Subsect. 5.3.

We also optimize our static move ordering with *attacker tie-breaking*. Assume a player is able to capture an opponent's piece using a pawn or a queen. Often it is better to take the piece using a pawn, in case the capturing piece is susceptible to a counter-capture. To assess the benefit of this optimization, we ran 1120 games of BRS^+ with and without it. The optimized version won $52.6\% \pm 3.0\%$ of the games. While the benefit is small, it still seems to have a positive effect on the move ordering, so it is included from here on. Finally, if two different moves capturing the same piece have the same attacker piece value, then values stored by the history heuristic (see below) are used as the final tie-breaker.

If there are no capture moves available from the static move ordering, then a dynamic move ordering may be applied. As is standard in search implementations, an iterative deepening version of BRS^+ is used with incrementally increasing depth d. Whenever a node is searched, the best move is stored at the node's entry in the transposition table. The first dynamic move ordering is then to consult the transposition table for this previously found best move. In addition, we try the killer-move heuristic and history heuristic. However, none of the dynamic move orderings seemed to improve performance significantly when enabled independently. In 560 games with a time limit of one second, enabling transposition tables leads to a win rate of $49.6\% \pm 4.2\%$. Similarly enabling killer moves and the history heuristic leads to a win rate of $50.8\% \pm 4.2\%$ and $51.7\% \pm 4.2\%$, respectively. We believe this is due to the instability of decision-making in games with more than two players, but more work is required for a concrete analysis on this point.

In the rest of the experiments, only the static move ordering is used with the attacker tie-breaking optimization.

5.3   Performance of BRS^+ vs. BRS, max^n, Paranoid

The first experiment compares the performance of BRS^+ with either paranoid or max^n move ordering when playing against one of the classic algorithms. Using a time limit of 1000 ms, each variant of BRS^+ is a player against each previous algorithm separately [2]. The results are shown in Table 3. The first thing to notice is that BRS^+ is winning in every instance. Clearly, the paranoid static move ordering is the better choice for BRS^+ as it is superior to the max^n ordering in all of our experiments. Finally, the performance of BRS^+ using the paranoid move ordering versus the classic algorithms increases significantly compared to the BRS from baseline experiments in Sect. 5.1, increasing by 11.1% (from 58.9% to 70%) against Paranoid and by 8.3% (from 56% to 64.3%) against max^n. This last results shows a clear benefit of BRS^+ over BRS when making direct comparisons to max^n and Paranoid.

Naturally, we also want know how BRS^+ performs when played in the multiplayer setting against the other two search techniques. First, we ran a similar

Table 3. Comparison of BRS variants vs. $\max^n$, Paranoid.

Static move ordering variant	$\max^n$	Paranoid	BRS
BRS^+ ($\max^n$ MO)	59.3% (±4.1%)	56.3% (±4.2%)	53.0% (±4.2%)
BRS^+ (Paranoid MO)	64.3% (±4.0%)	70.0% (±3.8%)	54.2% (±3.0%)
BRS	56.0% (±4.2%)	58.9% (±4.1%)	-

Table 4. BRS^+ vs. $\max^n$ vs. Paranoid.

Time (ms)	Games	BRS^+	$\max^n$	Paranoid
1000	720	49.4% ± 3.7%	17.6% ± 2.8%	33.1% ± 3.5%

Table 5. BRS^+ vs. BRS vs. $\max^n$ vs. Paranoid.

Time (ms)	Games	BRS^+	BRS	$\max^n$	Paranoid
1000	960	32.9% ± 3.0%	26.5% ± 2.8%	17.7% ± 2.5%	22.9% ± 2.7%
5000	960	35.4% ± 3.1%	22.7% ± 2.7%	17.4% ± 2.4%	24.5% ± 2.8%

Table 6. Average depths reached by search algorithms.

Time (ms)	Positions	BRS^+	$\max^n$	Paranoid
1000	200	6.105	3.075	4.115
5000	200	7.125	3.895	4.845

experiment to the baseline experiments in the three algorithm setting including BRS^+, Paranoid, and $\max^n$. The results are presented in Table 4. Again, the performance of BRS^+ is improved significantly compared to BRS, from 38.4% to 49.4% compared to results from Table 1. In addition, BRS^+ becomes the decisive winner by a margin of 16.3%.

Finally, we ran an experiment including all four algorithms. The results are shown in Table 5. Again, BRS^+ is the winner, beating the second place player roughly by a 5% gap. This gap more than doubles when the time limit is increased to five seconds. To confirm our expectation that BRS^+ is searching more deeply, we ran an additional experiment to compute the depth reached by each technique in the time allowed on a suite of 200 board positions. The results are presented in Table 6. These results show that, on average, BRS^+ reaches 1.99 to 2.28 ply more deeply than Paranoid for these time settings.

5.4   Rand-Top-k Move Ordering Experiment

Because the performance depends heavily on the move ordering, we also tried randomly choosing among the top k moves in the move ordering. This adds some variation in the moves considered by the opponents, which may lead to more robustness against bias in the move ordering. We compare the performance of

Table 7. Performance of BRS$^+$ with Rand-Top-k optimization.

Time (ms)	$k = 2$	$k = 3$	$k = 4$	$k = 5$	$k = 10$
1000	$51.6 \pm 2.1\%$	$53.3 \pm 2.1\%$	$51.2 \pm 2.1\%$	$52.2 \pm 2.1\%$	$52.3 \pm 2.1\%$
5000	$51.0 \pm 3.7\%$	$54.1 \pm 3.8\%$	$56.2 \pm 3.8\%$	$57.6 \pm 3.7\%$	$55.8 \pm 4.1\%$

Rand-Top-k directly by playing games with it enabled and disabled. The results are shown in Table 7. It seems that randomizing over the top k moves further improves the performance of BRS$^+$. These initial investigations suggest that the effect may not be smooth in the value of k at the lower time setting.

6 Conclusions and Future Research

In this paper, we introduced a new multi-player search technique, BRS$^+$, which can avoid invalid states and preserve the proper turn order during its search. BRS$^+$ expands asymptotically fewer nodes than the classic algorithms max^n and Paranoid as $d \rightarrow \infty$ leading to deeper searches. When BRS$^+$ uses move-ordering moves at opponent nodes that are not searched, its performance critically depends on the move ordering used. Through experiments in Four-Player Chess, we show that a paranoid static move ordering outperforms a static max^n move ordering in BRS$^+$. Finally, in all experiments the performance of BRS$^+$ is significantly higher than its predecessor, BRS.

For future work, we aim to compare our algorithms to the Comixer and MP-Mix algorithms mentioned in Sect. 3. In addition, we hope to extend the algorithm to the multi-player Monte-Carlo Tree Search [6,9] setting and apply it to other multi-player games such as Chinese Checkers, Blokus, and Focus as well as Hearts, which was not formerly playable by BRS.

Acknowledgments. We would like to thank Nathan Sturtevant for the suggestion of Rand-Top-k and Pim Nijssen for his help with multi-player search algorithms. This work is partially funded by the Netherlands Organisation for Scientific Research (NWO) in the framework of the project Go4Nature, grant number 612.000.938.

References

1. Campbell, M., Hoane Jr, A.J., Hsu, F.: Deep blue. Artif. Intell. **134**(1–2), 57–83 (2002)
2. Esser, M.: Best-reply search in multi-player chess. Master's thesis, Department of Knowledge Engineering, Maastricht University (2012)
3. Gras, M.: Multi-player search in the game of billabong. Master's thesis, Department of Knowledge Engineering, Maastricht University (2012)
4. Lorenz, U., Tscheuschner, T.: Player modeling, search algorithms and strategies in multi-player games. In: van den Herik, H.J., Hsu, S.-C., Hsu, T., Donkers, H.H.L.M.J. (eds.) CG 2005. LNCS, vol. 4250, pp. 210–224. Springer, Heidelberg (2006)

5. Luckhart, C.A., Irani, K.B.: An algorithmic solution of n-person games. In: AAAI'86, pp. 158–162 (1986)
6. Nijssen, J.A.M., Winands, M.H.M.: Search policies in multi-player games. ICGA J. **36**(1), 3–21 (2013)
7. Schadd, M.P.D., Winands, M.H.M.: Best reply search for multi-player games. IEEE Trans. Comput. Intell. AI Game **3**(1), 57–66 (2011)
8. Schaeffer, J., Burch, N., Björnsson, Y., Kishimoto, A., Müller, M., Lake, R., Lu, P., Sutphen, S.: Checkers is solved. Science **317**(5844), 1518–1522 (2007)
9. Sturtevant, N.R.: An analysis of UCT in multi-player games. ICGA J. **31**(4), 195–208 (2008)
10. Sturtevant, N.R., Korf, R.E.: On pruning techniques for multi-player games. In: Proceedings of the Seventeenth National Conference on Artificial Intelligence (AAAI 2000), pp. 201–207 (2000)
11. Tesauro, G.: Temporal difference learning and TD-Gammon. Commun. ACM **38**(3), 58–68 (1995)
12. Tscheuschner, T.: Four-person chess (Paderborn rules) (2005)
13. Zuckerman, I., Felner, A.: The MP-Mix algorithm: Dynamic search strategy selection in multiplayer adversarial search. IEEE Trans. Comput. Intell. AI Game **3**(4), 316–331 (2011)

Scalable Parallel DFPN Search

Jakub Pawlewicz[1] and Ryan B. Hayward[2($\boxtimes$)]

[1] Institute of Informatics, University of Warsaw, Warsaw, Poland
[2] Computing Science, University of Alberta, Edmonton, Canada
pan@mimuw.edu.pl, hayward@ualberta.ca

Abstract. We present Scalable Parallel Depth-First Proof Number Search, a new shared-memory parallel version of depth-first proof number search. Based on the serial DFPN $1 + \varepsilon$ method of Pawlewicz and Lew, SPDFPN searches effectively even as the transposition table becomes almost full, and so can solve large problems. To assign jobs to threads, SPDFPN uses proof and disproof numbers and two parameters. SPDFPN uses no domain-specific knowledge or heuristics, so it can be used in any domain. Our experiments show that SPDFPN scales well and performs well on hard problems.

We tested SPDFPN on problems from the game of Hex. On a 24-core machine and a 4.2-hour single-thread task, parallel efficiency ranges from 0.8 on 4 threads to 0.74 on 16 threads. SPDFPN solved all previously intractable 9×9 Hex opening moves; the hardest opening took 111 days. Also, in 63 days, it solved one 10×10 Hex opening move. This is the first time a computer or human has solved a 10×10 Hex opening move.

1 Introduction

Depth-First Proof-Number search is effective for solving problems — e.g., two-player games — that can be modeled with an and-or tree. It is especially effective on trees with non-uniform branching, and has been successful in checkers [1], shogi [2], tsume-shogi [3], Go [4–6], and Hex [7,8].

DFPN search often jumps around the tree. An effective parallel DFPN variant thus needs a shared *transposition table* (tt). The starting point for Scalable Parallel DFPN search, our shared-memory DFPN variant, is the serial $1+\varepsilon$-method [9], which is effective when the search space exceeds the available memory.

Other parallel variants based on proof numbers have been proposed. Nagai introduced proof-disproof search [10] and Kishimoto gave a parallel version [11]. Saffidine et al. proposed a job-level PN^2 [12] which is effective when the search space exceeds the available memory, although nodes can be recomputed many times. Kaneko achieves a parallel efficiency of 0.5 with 8 threads in tsume-shogi [2], whereas SPDFPN achieves 0.7 in Hex. Hex solvers tend to obtain better speedups than tsume-shogi solvers, perhaps because Hex has a larger branching factor and so a smaller node expansion rate.

Saito et al. introduced randomized parallel PNS [13]. Wu et al. [14] use PNS with virtual proof and disproof numbers and a time-intensive leaf initiation in a

H.J. van den Herik et al. (Eds.): CG 2013, LNCS 8427, pp. 138–150, 2014.
DOI: 10.1007/978-3-319-09165-5_12, © Springer International Publishing Switzerland 2014

tree that is small enough to stay in memory. SPDFPN is inspired by these virtual numbers, but is based on DFPN rather than PNS. It distributes work among threads using only proof and disproof numbers and thus needs no game-specific heuristic.

We tested SPDFPN on two sets of Hex problems, including all thirteen previously intractable 9×9 opening moves. See Fig. 1. Experiments show that our algorithm scales well up to at least 16 threads.

2   DFPN

Below we introduce DFPN in two steps, viz. by a brief description of proof-number search (Sect. 2.1), followed by a description of depth-first PN search (Sect. 2.2).

2.1   PN Search

Proof number search maintains a tree in which each node — corresponding to a game position — has a *proof number* (pn) and *disproof number* (dn). A node's (dis)proof number is the smallest number of leaves that, if all true (false), would make the node true (false). Thus for an or-node $p = \min_j p_j$ and $d = \sum_j d_j$, where p (d) is the node's (dis)proof number, and p_j (d_j) the j'th child's (dis)proof number. PN search relies on the existence of a most-proving node (mpn): this node's (dis)proof reduces either the proof or disproof number. PN search iteratively selects a most-proving leaf and expands it. PN search needs the whole search tree to be in memory. This restriction can be mitigated by taking a depth-first search approach and using a TT.

2.2   Depth-First PN Search

In computing (dis)proof numbers, only descendant (dis)proof values are needed. DFPN search exploits this property by postponing ancestor updates until the most-proving node is no longer in the current node's subtree. For each node, thresholds P and D are defined so that $p < P$ and $d < D$ if and only if there exists a most-proving node in the node's subtree. These thresholds are computed recursively using the following formula for an or-node. Children are indexed in non-decreasing proof number, e.g., $p_1(d_1)$ are the (dis)proof numbers of a child with smallest proof number. See [15] or [9] or the recent survey on game tree search using pns [16].

$$P_1 = \min(P, 1 + p_2), \qquad (1)$$
$$D_1 = D - (d - d_1). \qquad (2)$$

3   Transposition Table and DAGs

To this point we have described the process for computing (dis)proof numbers in trees. Many games allow transpositions, and so are modeled by *directed acyclic graphs* (dags) rather than trees. While more complicated variants of DFPN are available for dags [3], we find that in Hex, good results are obtained by treating the dag as a tree, as long as a tt is used.

3.1   Problems, Memory and the $1 + \varepsilon$ Method

Even with a tt, DFPN search behaves differently from the usual depth-first search, as the continued selection of the next mpn causes frequent switching among child branches. To force DFPN to switch branches less often, one can use the $1 + \varepsilon$ method [9], which simply enlarges the pn threshold in (1):

$$P_1 = \min(P, \lceil (1 + \varepsilon)p_2 \rceil).$$

A second problem arises in (d)pn calculations when dags are treated as trees: due to transpositions, (d)pns can be (exponentially) overcounted. Miscounting can lead to the mistaken selection of a node which is not most-proving. Techniques address this issue in games, such as tsume-shogi, with many transpositions [3]. We chose rather to address these issues by starting with the $1 + \varepsilon$ method, which reduces branch switching even if the search space is large. For our Hex experiments, possibly miscounting (dis)proof numbers did not cause difficulties. After trial and error, we set ε to 0.25.

4   Parallelization

The version of PN search we have described to this point behaves like a usual depth-first algorithm, spending significant periods of time in deep recursive calls. This allows for parallelization if the following five criteria, which motivate the design of SPDFPN, can be met.

(i) Different threads should call DFPN searches for different states.
(ii) A thread should not duplicate the work of another. (The threads share a tt, and different threads might explore different strategies for the same state.)
(iii) Assignment of states to threads should follow the natural order of PN search. (Assume that state A is assigned to thread α. If state B is likely to be the next state considered, we can assign B to another thread before α finishes.)
(iv) The assignment of states to threads should take little time.
(v) A thread should exit its search once other thread results render it unnecessary.

4.1   Measuring Work

We now show how we realize our criteria. Because of (v) we want to allow individual threads to halt and perhaps later resume. A straightforward attempt to parallelize DFPN that assigns a node to a processor, and has the processor return only when the subtree is solved, would not permit this.

To realize (v), we set a work threshold for a single DFPN call. The threshold ($MaxWorkPerJob$) must be small enough to allow even distribution among threads, but not so small that we lose (iv). If it is large enough we satisfy (iv) and can use more sophisticated methods for state assignment among threads. We need such a threshold, since (d)pns in a DFPN search can remain low and not reach any (d)pn thresholds for long periods of time, especially in sudden-death games.

Implementation. So how do we alter DFPN to incorporate the work threshold? In addition to the DFPN threshold parameters P and D, we introduce a new threshold parameter W, the maximum work that a thread should perform before halting. We define work as the number of calls to the DFPN function.

The $1 + \varepsilon$ method works best with advanced tt collision resolution. We use a method from computer chess (e.g., [17,18]) with $k = 4$: upon collision, search the next at most k cells for an empty location; if none is found, overwrite the location of which the job has performed the smallest amount of work. See [15,19,20] for other replacement or collection techniques.

SPDFPN pseudocode in Algorithms 1 and 2 uses these variables: c — array of child nodes; P, D — proof and disproof number thresholds; W — work threshold; n — node. A node has fields $.p$, $.d$ (proof and disproof numbers), $.w$ (work), $.j$ (index of last selected child). TTWRITE(n) writes results for node n in the tt. TTREAD(n, j) tries to read the jth child of node n; if this fails, it creates a child node with (d)pns each set to 1 and work set to 0.

Algorithm 1. DFPN Search

1: **function** DFPN(n, P, D, W)	9: **return** w_{local}
2: $w_{local} \leftarrow 1, n.w \leftarrow n.w + 1$	10: **if** $w_{local} \geq W$ **then**
3: **for each** child j of n **do**	11: **return** w_{local}
4: $c[j] \leftarrow$ TTREAD(n, j)	12: $j, P_j, D_j \leftarrow$ SELECT(n, c, P, D)
5: **loop**	13: $w_{child} \leftarrow$
6: PNUPDATE(n, c)	14: DFPN($c[j], P_j, D_j, W - w_{local}$)
7: TTWRITE(n)	15: $w_{local} \leftarrow w_{local} + w_{child}$
8: **if** $n.p \geq P \vee n.d \geq D$ **then**	16: $n.w \leftarrow n.w + w_{child}$

DFPN returns the amount of work done locally, i.e., in this call and all recursive calls. SELECT returns a child together with thresholds for recursive call. In the single-threaded version of the $1 + \varepsilon$ method, processing remains at a node n_1 until its pn p_1 exceeds the second-smallest pn p_2 (among siblings) by a ratio of $1 + \varepsilon$. Since we allow processes to be interrupted, it can be that upon resumption p_1 is larger than p_2 but smaller than $p_2(1 + \varepsilon)$. In this case, processing should resume at n_1 rather than its sibling. This requires adding to function SELECT the if-statement in lines 6–11.

5   Work Assignment

What are a thread's candidates for state assignment? PN-Search descends through a path from the root to a most-proving node. In DFPN this path is created by successive recursive calls and so is stored on a stack. However, DFPN search can remain deep in the search tree for long periods. Thus for a node of which the path is close to the most-proving node, DFPN can stay in a subtree of that node for some time. Such a node is a good candidate for a thread state assignment. But how deep on the subtree path should the assigned state be? It should be sufficiently deep that DFPN will stay in subtree of the node, not too shallow because of (v) and not too deep because of (iv). A good candidate is a node that is closest to the root and with past work performed below $MaxWorkPerJob$, because we expect that the total DFPN work for this node will be proportional to the total DFPN past work.

Once we assign a state A to a thread α, how should we assign a state B to another thread? If we follow the same procedure we would arrive at the same state. Instead, following Rémi Coulom (see [21, p. 64]), we temporarily assign a virtual win or loss to A until α finishes its search. This idea is also used by Job Level PN search, which achieves superlinear scalability, and fulfils (i), (ii) and (iii). See [14].

Algorithm 2. DFPN Search — utility functions for OR node

1: **procedure** PNUPDATE(n, c)
2:     $n.p \leftarrow \min\limits_{\text{child } j \text{ of } n} c[j].p$
3:     $n.d \leftarrow \sum\limits_{\text{child } j \text{ of } n} c[j].d$

4: **function** SELECT(n, c, P, D)
5:     $j_1 \leftarrow$ child with the smallest pn
6:     **if** $n.j$ is set and $n.j \neq j_1$ **then**
7:         ▷ Try continue with the same child
8:         $P_{n.j}, D_{n.j} \leftarrow$
9:             THRESHOLDS$(n, c, P, D, n.j, j_1)$

10:        **if** $c[n.j].p < P_{n.j}$ **then**
11:            **return** $n.j, P_{n.j}, D_{n.j}$
12:     $n.j \leftarrow j_1$
13:     $j_2 \leftarrow$ child with the second smallest pn
14:     $P_{j_1}, D_{j_1} \leftarrow$
15:         THRESHOLDS(n, c, P, D, j_1, j_2)
16:     **return** j_1, P_{j_1}, D_{j_1}
17: **function** THRESHOLDS(n, c, P, D, j_1, j_2)
18:     $P_{j_1} \leftarrow \min(P, \lceil (1 + \varepsilon) \cdot c[j_2].p \rceil)$,
19:     $D_{j_1} \leftarrow D - (n.d - c[j_1].d)$
20:     **return** P_{j_1}, D_{j_1}

5.1   Virtual Proof and Disproof Numbers

As we set the value of node A to virtual win or loss depending on its (d)pn, we must update other (d)pns along the path to the root. If we modify existing (d)pns, then other threads can reach a state with incorrect (d)pns via transposition. So the use of virtual win/loss requires the use of virtual (d)pns. An inaccurate assignment of a virtual win or loss will cause SPDFPN search to diverge from DFPN search, violating condition (iii). Our initial assignments of these virtual win/loss values are often accurate, as they are often made at nodes that have already been partly searched. The amount of previous search is determined by the threshold parameter $MaxWorkPerJob$. Virtual (d)pns can be stored efficiently (see below). In descending from the root towards a most-proving node in order to find a state assignment for a thread, we use virtual (d)pns if available, otherwise true (d)pns. After assigning a state A to a thread α, we update virtual (d)pns along the path to the root. Solving work then starts by calling DFPN on state A. This call uses true (d)pns. When the call returns, we reset the virtual win/loss back to a true (d)pn and update virtual (d)pns along the path to root. This completes an iteration of a thread loop, i.e., the thread now seeks its state assignment.

The entire phase of candidate-finding is guarded by a lock, so only a single thread can operate on virtual (d)pns at a time. A lock is released only when DFPN is called and solving work resumes.

Virtual (d)pns are kept in a virtual tt, which stores a node's virtual (d)pns and the number of threads assigned on the path to the root that contain that node. A node is added to the virtual tt at most as many times as the number of threads. This tt is easy to implement for dags. Each level (move) of the game corresponds to an array of which the size is at most the number of threads. Virtual tt operations are as follows.

- VTTADD(n): if the entry already exists, increment the counter.
- VTTREMOVE(n): decrement the counter; if 0, remove the entry, otherwise restore n's previous virtual (d)pdn.
- VTTREAD(n, j, n_j): return entry for n's jth child; if no such entry then initialize v(d)pns by returning the default n_j, which contains true (d)pns.

5.2 Finding a State Candidate for a Thread

Finding a candidate for state assignment can fail, as follows. Assume we are at node n. We have read its (d)pns from the tt. We must select a child to descend to, so we also read all of its children's (d)pns from the tt. Normally, the recursive formulas should give the (d)pns of n from those of its children. A child's (d)pns can be lost due to tt overwrites, but this is not a problem, as the (d)pns are recalculated when we descend to such child. The problem is that c was reached via transposition and work was done at c after the last update of n. Thus (d)pns can be stale, and an update can reveal that (d)pn thresholds were reached or even that n has been solved. So descending to a mpn via the usual rules is not sufficient. Instead, we recursively search as in DFPN, using virtual (d)pns whenever they exist, and stopping the search as soon as we find a state with the characteristic that the previous work performed is below a fixed threshold ($MaxWorkPerJob$). This search is performed by TRYRUNJOB, explained in Sect. 5.4.

5.3 Sharing Transposition Table

Threads share the tts, so we use multiple-reader/single-writer locks. Following [2], if a worker thread discovers immediately before writing that the node has (during the worker's processing) been solved by another thread, we do not overwrite the tt.

5.4 Implementation

Algorithm 3 shows the main scheme of SPDFPN, our parallel DFPN search. In the loop, each thread (1) calls TRYRUNJOB, (2) tries to find a candidate for state assignment, and (3) if successful then runs a job by calling DFPN on the assigned state. But first we need to update all virtual (d)pns, accessing them in nodes on the node-to-root path. We also need virtual (d)pns of children of each such node. So, we introduce a list v directed towards the root. Each list entry contains this data for the associated node: $v.n$ — a node with virtual (d)pns, $v.c$ — array of the node's children with (d)pns, $v.parent$ — refer to corresponding parent's v.

Algorithm 3. Parallel DFPN	
1: **procedure** PARALLELDFPN(*root*)	8: *job_done* ← false
2: **for** $i = 1, \ldots, \#$ of threads **do**	9: TRYRUNJOB(n, v, ∞, ∞)
3: spawn thread with call RUN(*root*)	10: UNLOCK(*job_lock*)
4: **procedure** RUN(*n*)	11: **if** *job_done* **then**
5: **while** n is not solved **do**	12: notify waiting threads
6: LOCK(*job_lock*)	13: **else**
7: $v.n \leftarrow n, v.parent \leftarrow$ null	14: wait

TRYRUNJOB is shown in Algorithm 4. It works as DFPN, but additionally calculates virtual (d)pns and stores them in list v. Once it finds a candidate — the condition in line 5 is true — virtual (d)pns are propagated upwards to the root and an actual job is run. Once this job is done the entire recursion ends, i.e., no more search is performed and virtual (d)pns are updated by VTTREMOVE calls.

5.5 Comparison to Kaneko's Algorithm

Kaneko parallelizes DFPN like this [2]: in an OR node's tt access, a child's pn is increased by the number of threads searching that child. Kaneko calls this augmented value a virtual pn[1], discouraging — but not preventing — the search from repeatedly selecting the same child. In our experiments we observe that, near tree-top, sibling pns vary, whereas near tree-bottom, they are all small and so similar. In the former case, Kaneko's algorithm is likely always to select the same child for search.

In contrast, in our approach, by setting the value of *MaxWorkPerJob*, we implicitly control how deep in the tree the diversion of thread selections should occur. Moreover, our threads always work on different subtrees.

6 Experiments

We implemented SPDFPN for Hex on the open-source Hex repository BENZENE [22], which in turn is built on the open-source game-independent

---

[1] Do not confuse Kaneko's augmented proof/disproof values with our definition of virtual pn.

framework FUEGO [23]. BENZENE uses Focussed DFPN search [8,24] (FDFPN), which employs an evaluation function to sort a node's children, and then focuses the search on a fraction of the most-promising children. The size of the search window is given by $\lceil b + f \times \#\text{active children} \rceil$, so new children that can enter the window as siblings are proved to be a loss. We use $b = 0$ and $f = 0.25$. FDFPN search maintains the usual correctness properties of PN search. We used FDFPN because it is embedded in BENZENE's DFPN; our use of this DFPN variant does not diminish the generality of SPDFPN.

Before starting our experiments, we improved BENZENE's virtual connection engine and solver. The resulting implementation performs typically 2 to 10 times faster than the previous version on similar hardware [7].

We tested SPDFPN on two sets of Hex problems: Suite 1, the thirteen previously intractable 9×9 opening moves plus the (previously intractable) centremost 10×10 opening move; and Suite 2, the eight hardest 8×8 opening moves plus eight positions from the 2011 Olympiad Hex competition [25].

Algorithm 4. Try find a candidate and run a job

```
 1: function TRYRUNJOB(n, v, P, D)
 2:     if v.n.p ≥ P ∨ v.n.d ≥ D then
 3:         return 0
 4:     w_local ← 0
 5:     if n.w < MaxWorkPerJob then
 6:         if n.p ≤ n.d then
                                          ▷ Virtual win
 7:             v.n.p ← 0, v.n.d ← ∞
 8:         else
                                          ▷ Virtual loss
 9:             v.n.p ← ∞, v.n.d ← 0
10:         UPDATEVIRTUALS(v)
11:         UNLOCK(job_lock)
                                          ▷ Candidate is found
                                          ▷ Actual job is run here
12:         w_local ←
13:             DFPN(n, P, D, MaxWorkPerJob)
14:         LOCK(job_lock)
15:         job_done ← true
16:         v.n.p ← n.p, v.n.d ← n.d
17:         VTTREMOVE(v.n)
18:         return w_local
19:     for each child j of n do
20:         c[j] ← TTREAD(n, j)
21:         v.c[j] ← VTTREAD(n, j, c[j])
22:     loop
23:         PNUPDATE(n, c)
24:         PNUPDATE(v.n, v.c)
25:         TTWRITE(n)
26:         if job_done then
27:             VTTREMOVE(v.n)
28:             return w_local
29:         if v.n.p ≥ P ∨ v.n.d ≥ D then
30:             return w_local
31:         j, P_j, D_j ←
32:             SELECT(v.n, v.c, P, D)
33:         v_child.n ← v.c[j], v_child.parent ← v
34:         w_child ←
35:             TRYRUNJOB(c[j], v_child, P_j, D_j)
36:         w_local ← w_local + w_child
37:         n.w ← n.w + w_child
38: procedure UPDATEVIRTUALS(v)
39:     VTTADD(v.n)
40:     while v.parent is not null do
41:         v ← v.parent
42:         PNUPDATE(v.n, v.c)
43:         VTTADD(v.n)
```

6.1   Previously Intractable 9×9 and 10×10 Hex Openings

Suite 1 tests the limits of SPDFPN on 24 threads of a hyperthreaded 12-core Intel Xeon 2.93 GHz with 48 Gbyte RAM and 8 threads of an 8-core Intel Xeon

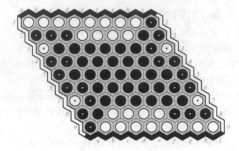 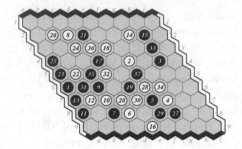

Fig. 1. Newly solved 9 × 9 opening values (dots), winner if black opens there.

Fig. 2. PV of a6, the hardest 9 × 9 opening.

2.8 GHz with 32 Gbyte RAM. We used a TT with sizes varying from 2^{27} to 2^{28} entries depending on machine and stage of a search. Here, in TRYRUNJOB, in addition to a TT we used a database storage capable of handling simple board isomorphism (180 degree rotation). For more difficult openings we gradually raised the value of ε from 0.25 up to 0.5 in order to reduce the number of TT lookup failures. See Fig. 1.

Table 1. Times (days:hrs:mins:secs) and threads for newly solved 9 × 9 and 10 × 10 openings.

opening	#threads	time	winner	opening	#threads	time	winner
a2	8/24	68d09:40:18	black	b2	8	53d15:18:21	black
a3	8	80d08:37:34	white	b4	8	29d23:53:14	black
a4	8	33d14:06:03	black	b6	8	1d21:52:28	black
a5	8	65d04:14:52	black	b7	8	4d17:19:13	black
a6	24	110d14:35:06	black	c2	24	1d08:42:57	black
a7	24	4d08:56:03	white	i1	24	6d00:51:25	black
a8	24	6d14:21:30	black	10x10:f5	24	63d20:44:30	black

Table 1 shows the number of threads used and the approximate running times. Due to occasional machine shutdown, e.g., power failure, some runs were restarted several times from database and tt backups; for these runs the running times are cumulative estimates based on logs. As an indication of achieved speedup on this problem suite, the previous algorithm with 8 threads failed to solve any of these openings after 480 h[2], whereas SPDFPN with 24 threads solves c2 in under 33 h. On the 9 × 9 board a6 was the hardest opening. The behavior of SPDFPN (with 24 threads) on this problem was different from all other configurations: in SPDFPN (24), the main lines of play were extremely balanced, and the winner unclear, until deep into the search. See Fig. 2. Although the search

---

[2] Private communication with Broderick Arneson.

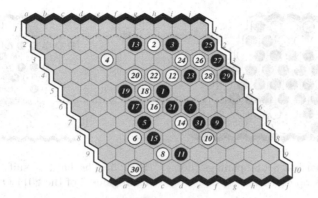

Fig. 3. Principle variation of f5, the first-ever solved 10×10 opening. Black wins.

space was around 100,000 times larger than the size of TT, SPDFPN showed continuous progress. In the previously strongest Hex solver the search often gets stuck whenever the search space is this much larger than the TT [8,22]. In this research we were able to solve a 10×10 opening. The principal variant was: f5. Black wins. See Table 1 and Fig. 3.

6.2   8 × 8 and Olympiad Hex Problems

Suite 2 measures the parallel efficiency of our algorithm. On a 24-core Intel Xeon 2.4 GHz with 64 Gbyte RAM, we used 16 threads (the others were in use). We used a tt with 2^{24} entries, which was more than sufficient. We picked moderate problems: challenging but still tractable for a single thread. This suite consists of eight (hardest) 8×8 openings and eight 11×11 positions from the 2011 ICGA Olympiad found by starting with the final position and proceeding backwards to a moderate position. See Figs. 4 and 5.

6.3   Scalability

In this experiment we measured scalability, or parallel efficiency, — the average[3] speedup ratio over serial version — for positions from suite 2. We ran our algorithm over all instances two times. For run 1, on board size 8×8 (11×11), the default value of *MaxWorkPerJob* was 100 (20). For run 2, *MaxWorkPerJob* was 500 (100). In many domains, small values such as these can yield lower scalability due to thread management overhead. However, our Hex solver spends a large fraction of the time on VC engine computations, so for this solver these small values of *MaxWorkPerJob* are suitable. Our algorithm scales well on up to 16 threads. Figure 6 shows the scalability from run 1 (9.4 with 16 threads, or .59), and from the six hardest problems — three each from 8×8 and 11×11 — from

---

[3] As usual when measuring a ratio (here, speedup), we use geometric mean for averaging.

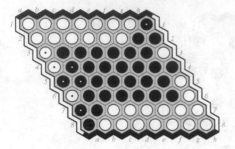

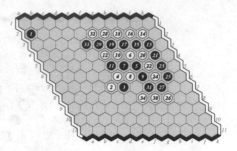

Fig. 4. The hardest 8×8 openings (dots), winner if black opens there.

Fig. 5. The hardest suite 2 position, from game 7 of the 2011 Olympiad.

Table 2. Scalability.

	Run 1			Run 2		
n	f_t^1	f_s^1	e_1	f_t^2	f_s^2	e_2
1	1.000	1.000	1.000	1.000	1.000	1.000
2	0.981	1.051	1.940	1.033	1.095	1.768
4	1.071	1.094	3.414	1.041	1.158	3.318
8	1.111	1.124	6.405	1.072	1.268	5.885
12	1.098	1.398	7.816	1.008	1.319	9.028
16	1.219	1.401	9.368	1.091	1.245	11.780

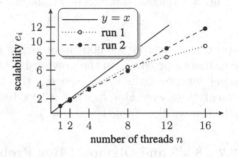

Fig. 6. Scalability.

run 2 (e.g., 11.8 with 16 threads, or .74). Table 2 shows how time is lost due to parallelization. For each run $i = 1, 2$, f_t^i denotes the average fraction of time lost due to multi-threading overhead (shared tt access, hardware overhead) and f_s^i denotes the average fraction of time lost due to extra leaf expands (extra states searched).

These values are computed as follows. Let s_1 and s_n be the number of leaf expands (number of states for which the VC engine was used) for the serial and n-thread runs respectively. Then $f_s^i = s_n/s_1$. Let t_1 and t_n be the actual running times for the serial and n-thread runs. If there is no multi-threading overhead then the expected time of the n threads run is $E_t = t_1 s_n/s_1/n$, so $f_t^i = t_n/E_t$. Thus $t_n = f_t f_s t_1/n$, so the scalability e_i is $n/(f_t^i f_s^i)$.

7   Conclusions

We have introduced SPDFPN, a parallel version of depth-first proof number search that scales well. We tested our algorithm on two suites of Hex problems, in the process of solving all thirteen previously intractable 9×9 openings and the first-ever solution to a 10×10 opening. Our experiments showed a speedup of .74, namely 11.8 on 16 threads. Our algorithm is general and game-independent, and so should be equally effective on any problem that can be modeled by and-or trees. It would be of interest to see whether the SPDFPN speedups we achieved in Hex can be achieved in other domains, and to compare and contrast SPDFPN to Kaneko's parallel DFPN.

Acknowledgements. We thank Broderick Arneson, Yngvi Björnsson, Phil Henderson, Aja Huang, Timo Ewalds, Martin Müller, and the referees for their feedback. We thank Martin for generously loaning the use of his computing cluster for our experiments.

References

1. Schaeffer, J., Burch, N., Björnsson, Y., Kishimoto, A., Müller, M., Lake, R., Lu, P., Sutphen, S.: Checkers is solved. Science **317**, 1518–1522 (2007)
2. Kaneko, T.: Parallel depth first proof number search. In: Proceedings of the AAAI-10, pp. 95–100 (2010)
3. Kishimoto, A.: Dealing with infinite loops, underestimation, and overestimation of depth-first proof-number search. In: Proceedings of the AAAI-10, pp. 108–113 (2010)
4. Kishimoto, A., Müller, M.: A solution to the ghi problem for depth-first proof-number search. Inf. Sci. **175**, 296–314 (2005)
5. Kishimoto, A., Müller, M.: About the completeness of depth-first proof-number search. In: van den Herik, H.J., Xu, X., Ma, Z., Winands, M.H.M. (eds.) CG 2008. LNCS, vol. 5131, pp. 146–156. Springer, Heidelberg (2008)
6. Yoshizoe, K., Kishimoto, A., Müller, M.: Lambda depth-first proof number search and its appplication to go. In: Proceedings of the IJCAI-07, pp. 2404–2409 (2007)
7. Henderson, P., Arneson, B., Hayward, R.: Solving 8 × 8 Hex. In: Proceedings of the IJCAI-09, pp. 505–510 (2009)
8. Arneson, B., Hayward, R.B., Henderson, P.: Solving hex: beyond humans. In: van den Herik, H.J., Iida, H., Plaat, A. (eds.) CG 2010. LNCS, vol. 6515, pp. 1–10. Springer, Heidelberg (2011)
9. Pawlewicz, J., Lew, Ł.: Improving depth-first PN-search: 1 + ε trick. In: van den Herik, H.J., Ciancarini, P., Donkers, H.H.L.M.J. (eds.) CG 2006. LNCS, vol. 4630, pp. 160–171. Springer, Heidelberg (2007)
10. Nagai, A.: A new AND/OR tree search algorithm using proof number and disproof number. In: Proceeding of Complex Games Lab Workshop, Tsukuba, ETL, pp. 40–45 (1998)
11. Kishimoto, A.: Parallel AND/OR tree search based on proof and disproof numbers. In: 5th Games Programming Workshop. IPSJ Symposium Series, vol. 99, pp. 24–30 (1999)
12. Saffidine, A., Jouandeau, N., Cazenave, T.: Solving BREAKTHROUGH with race patterns and job-level proof number search. In: van den Herik, H.J., Plaat, A. (eds.) ACG 2011. LNCS, vol. 7168, pp. 196–207. Springer, Heidelberg (2012)
13. Saito, J.-T., Winands, M.H.M., van den Herik, H.J.: Randomized parallel proof-number search. In: van den Herik, H.J., Spronck, P. (eds.) ACG 2009. LNCS, vol. 6048, pp. 75–87. Springer, Heidelberg (2010)
14. Wu, I.-C., Lin, H.-H., Lin, P.-H., Sun, D.-J., Chan, Y.-C., Chen, B.-T.: Job-level proof-number search for connect6. In: van den Herik, H.J., Iida, H., Plaat, A. (eds.) CG 2010. LNCS, vol. 6515, pp. 11–22. Springer, Heidelberg (2011)

15. Nagai, A.: Df-pn Algorithm for Searching AND/OR Trees and its Applications. Ph.D. thesis, University of Tokyo, Japan (2002)
16. Kishimoto, A., Winands, M., Müller, M., Saito, J.T.: Game-tree search using proof numbers: the first twenty years. ICGA **35**, 131–156 (2012)
17. Letouzey, F.: Fruit (2004–2013). http://www.fruitchess.com/
18. Romstad, T.: Stockfish (2008–2013). http://stockfishchess.org/
19. Breuker, D., Uiterwijk, J., den Herik, H.: Replacement schemes and two-level tables. ICGA **19**, 175–180 (1996)
20. Nagai, A.: A new depth-first-search algorithm for and/or tree. Master's thesis, University of Tokyo, Japan (1999)
21. Chaslot, G.M.J.-B., Winands, M.H.M., van den Herik, H.J.: Parallel Monte-Carlo tree search. In: van den Herik, H.J., Xu, X., Ma, Z., Winands, M.H.M. (eds.) CG 2008. LNCS, vol. 5131, pp. 60–71. Springer, Heidelberg (2008)
22. Arneson, B., Henderson, P., Hayward, R.B.: Benzene (2009–2012). http://benzene. sourceforge.net/
23. Enzenberger, M., Müller, M., Arneson, B., Segal, R., Xie, F., Huang, A.: Fuego (2007–2012). http://fuego.sourceforge.net/
24. Henderson, P.: Playing and solving Hex. Ph.D. thesis, University of Alberta (2010). http://webdocs.cs.ualberta.ca/~hayward/theses/ph.pdf
25. Hayward, R.B.: 2011 ICGA Computer Games Olympiad Hex Competition Report (2011). http://webdocs.cs.ualberta.ca/~hayward/papers/rptTilburg.pdf

A Quantitative Study of 2 × 4 Chinese Dark Chess

Hung-Jui Chang and Tsan-sheng Hsu[✉]

Institute of Information Science, Academia Sinica, Taipei 115, Taiwan
{chj,tshsu}@iis.sinica.edu.tw

Abstract. In this paper, we study Chinese dark chess (CDC), a popular 2-player imperfect information game that is a variation of Chinese chess played on a 2×4 game board. The 2×4 version is solved by computing the exact value of each board position for all possible fair piece combinations. The results of the experiments demonstrate that the initial arrangement of the pieces and the place to reveal the first piece are the most important factors to affect the outcome of a game.

1 Introduction

Chinese dark chess (CDC) is a popular version of Banqi [1], a variation of *Chinese Chess* (CC). It uses the same set of pieces as CC, but only requires half of the game board. There are 32 pieces: 16 red pieces and 16 black pieces; and each color has one king (K or k), two guards (G or g), two ministers (M or m), two rocks (R or r), two knights (N or n), two cannons (C or c) and five pawns (P or p). In this paper, the red pieces and black pieces are denoted by upper case characters and lower case characters respectively. Each piece has two sides, a dark side and a revealed side. A piece's type is crafted on the revealed side and the dark sides of all the pieces are identical. Moreover, we say a piece is "dark" if its dark side is facing up and it is "revealed" if its revealed side is facing up. In contrast to CC, each CDC piece is put in a cell, not at an intersection. At the beginning of a game, all the pieces are "dark". Thus the players do not know the type of each piece. The first player flips one of the dark pieces and then "owns" the color of that piece, meaning that the second player "owns" the other color. In each turn, a player can (1) flip one of the dark pieces to reveal its identity, or (2) move one of his own revealed pieces. A reveal piece can be moved to an adjacent empty cell or be used to capture a lower-ranked or same-ranked revealed opponent piece in the adjacent cell. There is a special capture rule for cannons which will be described later. A player loses the game if he cannot make any legal move. The pieces can be ranked in four ways: (1) $K > G > M > R > N > C$, i.e., king, guard, minister, rook, knight and cannon; (2) $G, M, R, N > P$ and $P > K$, i.e., the pawn's rank is less than that of the guard, minister, rock and knight, but higher than that of the king; (3) there is no order between the cannon and the pawn; and (4) except for the cannon, every piece is of the same order as itself. The cannon can make a special *capture* move in the following circumstances.

H.J. van den Herik et al. (Eds.): CG 2013, LNCS 8427, pp. 151–162, 2014.
DOI: 10.1007/978-3-319-09165-5_13, © Springer International Publishing Switzerland 2014

On a straight line, if there is exactly one piece, dark or revealed between a cannon and a revealed opponent piece, the cannon can capture this revealed opponent piece, no matter its rank, by *jumping* to its location.

A number of previous work on CDC [1,2,10] have been mostly worked on a particular problem, namely, the uncertainty of a flip move. Because a player does not know the type and color of a dark piece until it is revealed, unless the remaining pieces are of the same type, the game is nondeterministic; and the probability of an event occurring will change over the game. Several games involve probability behaviors, e.g., Backgammon, Bridge, and Mahjong. However, in contrast to Backgammon, the probability of a particular event to occur in CDC changes during the game [8] but in Backgammon it does not. This dynamic behavior of CDC is more like that in Bridge [7] and Mahjong [6]. However, these are normally games with four players involved. As a research topic, CDC contains the following important properties: (1) it is a 2-player game; (2) it is a stochastic game and the probability of events will change during the game; and (3) at the endgame stage, when all pieces are revealed it becomes a perfect information game.

Our objective is to determine the dynamics of the occurring distributions over the course of playing a game. Currently, it is computationally impossible to solve the 4×8 full version completely, so that we study a smaller version of the game to gain some insight. Furthermore, we create a new variation of CDC, called the *oracle version*. Unlike the normal version, the players in the oracle version know the type of each dark pieces; hence, the game becomes a perfect information game. For the normal version and the oracle version, we also study them using two different opening rules. The first is called the normal opening rule and the second is called *color assigned* rule. If the color assigned rule is used, the first player's pieces are always red, that is, the first player may reveal one of the opponent's pieces in his[1] first move. The normal opening rule is the first player owns the color of the first revealed piece which is the default rule.

By solving the 2×4 case completely, we hope to gain insight into the game's behavior which may be used to solve the original version. We reduce the size of the game board from 4×8 to 2×4; and hence the number of cells is 8 instead of 32, which means that each player only holds 4 pieces. To ensure fairness, the players have an identical set of pieces. We evaluate all 24 fair combinations in our experiments. Figure 1(a) shows the initial board positions of 2×4 CDC. Each position on the board has a unique ID between 0 and 7. Figure 1(b) shows the relationship between each ID and its coordinate.

The remainder of this paper is organized as follows. In Sect. 2 we introduce the indexing function of position and the method used to solve the game completely. In Sect. 3, we describe the experimental environment and show the experimental results. In Sect. 4 we consider some implementation issue and discuss the experimental results. Section 5 contains some concluding remarks.

---

[1] For brevity, we use 'he' and 'his' whenever 'he or she' and 'his or her' are meant.

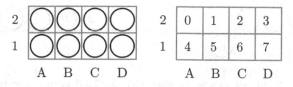

(a) A beginning board.   (b) The cell ID of each position

Fig. 1. The initial board

2   Methods

In this section, we describe the methods used to solve 2 × 4 CDC. First, we solve the oracle version of 2 × 4 CDC. Then, we use the information derived from the oracle version to solve the normal version. Subversions using the color assigned rule and the normal opening rule are generated at the same time because a board position has the same value in both of the two subversions except for the initial board positions.

2.1   Oracle Version of 2 × 4 CDC

In the oracle version of 2 × 4 CDC, the type of a dark piece in a cell is known by both players even it is dark. Thus it is a perfect information game. To solve this version completely, we use a bottom up approach that is similar to the retrograde analysis method used in [9].

Indexing Function. Before explaining how we solve the 2×4 CDC, we describe the indexing function used. The index of a board position can be divided into two parts: the position part and the revealed-dark part. For each combination, we use a unique sequence to list all the pieces in the combination. In this paper, we use a substring of "KGMRNCPkgmrncp" to list the pieces in a combination. For example, a combination of a red king, a red minister, a black guard and a black pawn would be written as "KMgp". To record a given board position in the oracle version of 2 × 4 CDC, we simply record each piece's position in the order given in the substring.

Figure 2(a) shows an example of the board in the oracle version of 2×4 CDC. There is a red king at D-1 with a cell index 7, a red minister at B-2 with a cell index 1, a black guard at C-2 with a cell index 2 and a black pawn at A-2 with a cell index 0. The red king and the black pawn are revealed, and the red guard and the black minister are dark. In this case, the position is recorded as (7, 1, 2, 0), and the status of pieces being revealed or dark is recorded as (0, 1, 1, 0) where 0 means revealed and 1 means dark or unrevealed. When a "d" is prefixed for a piece, it means a dark piece. For example, in Fig. 2(a), dM inside cell B-2 means it is a dark Minister.

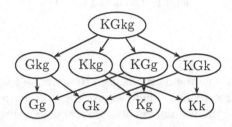

(a) (7,1,2,0,0,1,1,0) in an oracle version game   (b) (2,35,3,3,1) in a normal version game

Fig. 2. The board position of "KGgp" in the oracle version and normal version.

Fig. 3. The dependency graph of "KGkg".

Since there are only eight cells, the status of a board with n pieces can be recorded using $4n$ bits, where $3n$ of the bits denote the positions and the last n bits denote the status of pieces being revealed or dark. For example, (7, 1, 2, 0, 0, 1, 1, 0) means an index of $(7 \times 8^3 + 1 \times 8^2 + 2 \times 8^1 + 0 \times 8^0) \times 2^4 + (0 \times 2^3 + 1 \times 2^2 + 1 \times 2^1 + 0 \times 2^0) = 56827$. We will explain the meaning of (2, 35, 3, 3, 1) for the normal version in Sect. 2.2.

When using our indexing function, we make two assumptions. First, some position codes do not represent a valid board position because they may refer to having more than one piece in the same cell. Those board positions are invalid. Second, when we solve 2×4 CDC, the board positions of red player's turn and the board positions of black player's turn are recorded separately. Hence, the information about the players' turn is not recorded in the indexing function.

Bottom-Up Methods. We use a bottom up method to solve the oracle version 2×4 CDC. For a given combination of pieces, all the proper subsets will be generated before the given one can be generated. For example, "KGkg" depends on "KGk", "KGg", "Kkg" and "Gkg". Figure 3 shows all the dependency relations of "KGkg". Note that although "Gg" is a proper subset of "Gkg" and "KGg"; however, "Gkg" and "KGg" cannot be transferred into "Gg" since the king can not be captured by the opponent.

For each combination of pieces, we determine the values of all positions as follows. First, we mark the valid positions as "Unknown" and the invalid positions as "Invalid". Then we update each valid positions as follows: (1) if the position has a winning child, it is updated as a winning position; or (2) if all the children are losing, the positions is marked as losing. Remaining positions with "Unknown" after a round of updates without making any further progress are set as "Draws".

2.2 Normal Version of 2 × 4 CDC

In the normal version of 2 × 4 CDC, a dark piece's type is unknown until it is revealed or all unrevealed pieces are of the same type. For each dark piece on the board, the players only know what types they can possible be. Hence, the value of each board position is recorded as a probability distribution with 3 possible outcomes: win, draw or loss. Due to this nature, we use the expected value of the difference between the winning probability and the losing probability to be the value of a board position. We also redesign the indexing function to avoid ambiguity in representing a board position.

New Indexing Function. In the normal version of 2 × 4 CDC, if we use the same indexing function as the oracle version, then there will be a problem. Because the type of a dark piece is unknown before it is revealed, ambiguity arises if there is more than one dark piece. For example, in the normal version, the board positions in Figure 2(a) are treated as the board positions in Fig. 2(b); that is, the type of dark pieces are undetermined yet, so the dark pieces are treated as the same. As the positions of the red minister and black guard are unknown in Fig. 2(b), the indexes $(7, 1, 2, 0, 0, 1, 1, 0)$ and $(7, 2, 1, 0, 0, 1, 1, 0)$ may refer to the same position.

To overcome this problem, we redesign the indexing function. A board position with a given combinations of pieces is recorded by the following five attributes: (1) the number of revealed pieces on the board; (2) the number of combinations of chosen cells which are occupied by the pieces from all cells; (3) the number of combinations of chosen cells which are occupied by the revealed pieces from the occupied cells; (4) the number of combinations of chosen revealed pieces from the combination of pieces; and (5) the number of permutations of ordering revealed pieces according to their occupied cell ID's. There is a 1-1 mapping between the i-th combination (or i-th permutation) and the integers. We use the method proposed by Knuth [4] to calculate the i-th combination of an ordered set and the i-th permutation of an ordered set. The range and notation of each attribute are listed in Table 1.

In Fig. 2(b), the number of revealed pieces is 2, and the cells occupied by the pieces are 0, 1, 2, and 7; hence the combination order equals to $\binom{0}{1} + \binom{1}{2} + \binom{2}{3} + \binom{7}{4} = 35$. The revealed pieces occupy cells 0 and 7, which are the first and fourth occupied cells; therefore, the combination order equals to $\binom{0}{1} + \binom{3}{2} = 3$. The red king and the black pawn are the first piece and the fourth piece, respectively, in the "KMgp" combination, so the combination order equals to $\binom{0}{1} + \binom{3}{2} = 3$. The cell ID of the red king is 7 and the cell ID of the black pawn is 0, so the order of

Table 1. The attributes of a board position.

	Description	From	To
TPIECE	The total number of pieces	1	8
LPIECE	The total number of revealed pieces	0	LPIECE
BLOCKID	The i-th subset of all cells with size TPIECE	0	$\binom{8}{TPIECE} - 1$
LBLOCKID	The i-th subset of occupied cell with size LPIECE	0	$\binom{TPIECE}{LPIECE} - 1$
LSET	The i-th subset of combination of pieces with size LPIECE	0	$\binom{TPIECE}{LPIECE} - 1$
LPER	The i-th permutation of revealed pieces with size LPIECE	0	$LPIECE! - 1$

revealed pieces according to the cell ID is "pK", which is the second permutation of the set {K, p}. This means its index is 1, So the index in Fig. 2(b) is (2, 35, 3, 3, 1).

Similar to the oracle version, the values of the red turns and the black turns are saved separately, so the information about the turns is not included in the index function. Because there is a 1-1 mapping between i-th combination and integers as well as a 1-1 mapping between i-th permutation and integers, each index represents a unique valid board position. Hence, there is no ambiguity.

Top-Down Method. The values of board positions in the normal version of the game are represented by expected values, instead of being exact, namely the value of each board position is an real number ranging from -1 to 1, that is 100 % loss and 100 % win. In this case, the value of a winning board position is 1 and that of a losing board is -1. The update strategy used in the oracle version can not be used on solving the normal version as most of the board positions having a value between -1 and 1 rather than equaling to -1 or 1. In this case, instead of using a bottom-up method, we use a top-down method to determine the value of each board position.

For each combination of pieces, we determine the value of the board positions from the case of having no dark piece to the case of all pieces are dark. Because flipping a dark piece is a non-reversible move, the number of dark pieces can not increase during the course of a game. Hence, the values of board positions consisting k dark pieces depend on the values of board positions consisting of $k-1$ dark pieces. When the number of dark pieces is one or zero, the normal version and the oracle version of game become the same, which means the former can use the values of the latter directly. To determine the values of board positions that have more than one kinds of dark pieces, we use an iterative deepening depth-first search strategy [5].

3   Experiments

In this section we describe the experimental setting and then discuss the experimental results.

Table 2. Statistical results of all the initial positions in the oracle version of the game.

Index	Pieces combination	Color assigned			Normal opening		
		Win	Draw	Loss	Win	Draw	Loss
1	CCPP v.s. ccpp	16896	14784	8640	12928	15552	11840
2	CPPP v.s. cppp	14400	20448	5472	9360	19872	11088
3	GCCP v.s. gccp	20368	5472	14480	18688	5504	16128
4	GCPP v.s. gcpp	21440	7040	11840	20112	7216	12992
5	GGCC v.s. ggcc	21280	5824	13216	19296	5888	15136
6	GGCP v.s. ggcp	22880	7024	10416	21712	6800	11808
7	GGMC v.s. ggmc	22496	8240	9584	21696	8176	10448
8	GGMM v.s. ggmm	17664	10304	12352	16832	10496	12992
9	GGMR v.s. ggmr	18624	11120	10576	18096	11216	11008
10	GMCP v.s. gmcp	20640	8752	10928	19532	8704	12084
11	GPPP v.s. gppp	16560	11952	11808	14256	12960	13104
12	KCCP v.s. kccp	17536	10736	12048	15696	10944	13680
13	KCPP v.s. kcpp	17824	13456	9040	16496	13120	10704
14	KGCC v.s. kgcc	19552	7568	13200	18416	7408	14496
15	KGCP v.s. kgcp	19872	8024	12424	18740	7932	13648
16	KGGC v.s. kggc	20832	10064	9424	19360	9888	11072
17	KGGM v.s. kggm	17328	12640	10352	15632	13280	11408
18	KGGP v.s. kggp	15984	8400	15936	14592	8720	17008
19	KGMC v.s. kgmc	19596	10900	9824	18724	10760	10836
20	KGMM v.s. kgmm	16224	13344	10752	14992	13680	11648
21	KGMP v.s. kgmp	16764	8740	14816	15584	8952	15784
22	KGMR v.s. kgmr	16820	13256	10244	15844	13492	10984
23	KPPP v.s. kppp	16272	7344	16704	15120	6768	18432
24	PPPP v.s. pppp	4608	0	35712	2304	0	38016

3.1   Experimental Setting

In the experiments, we use 24 fair combinations of pieces as shown in Tables 2 and 3. We then conduct experiments on the oracle version and the normal version of the game. For the oracle version, we show the number of wins, losses, and draws for that initial positions of each combination of pieces using the normal opening rule and the color assigned rule. For the normal version, we show the expected difference between the number of winning events and the number of losing events. That is, we set the values of a win, draw and loss at 1, 0 and −1, respectively. Note that some of the results may have the same expected difference, but the original distribution is not the same; for example, if the total number of events is 2, then 1 win and 1 loss (1W1L) and 2 draws (2D) have the same expected difference of 0. We ran the experiments on a server with an Intel Xeon CPU E5-2690 2.90 GHz and 96 GB DDRIII 1333 memory.

3.2   Experimental Results of the Oracle Version

The experimental results of the oracle version of the game are listed in Table 2. Column 1 shows the combinations of pieces; columns 2–4 show the experimental results using the color assigned rule and columns 5–7 show the experimental results using the normal opening rule. There are eight pieces on the board, so the total number of initial positions in 2×4 Chinese dark chess is $8! = 40,320$. The results show that, when the color assigned rule is used, most of the cases favor the first player. The only exceptions are "KPPPkppp" and "PPPPpppp". Comparison of the results derived from the two different opening rules shows that, for all combinations, the normal opening version has less winning positions and more losing positions than the color assigned version.

3.3   Experimental Results of the Normal Version

In the normal version of game, the pieces are shuffled and the game board is symmetric. So there are only two choices for the first move, namely, A-2 (cell 0) and B-2 (cell 1). Therefore, we can reduce the choice of where to make the first move to two. Furthermore, each first move is followed by seven possible secondary moves. As with the oracle version, the total number of all possible cases is $40,320$. Table 3 lists the experimental results of the normal version of the game. Columns 1 and 2 show the index and the combination of pieces, columns 3–6 show the experimental result if the color assigned rule is used. Columns 3 and 4 show the expected value and the standard variation when the first flip cell is A-2 (cell 0); Columns 5 and 6 show the expected value and the standard variation when the first flip cell is B-2 (cell 1). Columns 7–10 show the experimental result if the normal opening rule is used and the format is the same as the experimental results if the color assigned rule is used. The results show that the first player is disadvantaged under most combinations of pieces. The only exceptions are "KGCPkgcp", "KGGCkggc" and "GCPPgcpp". The standard variation is 0.9; that is, although the expected result favors the second player, the second player does not have a sure win.

Table 4 lists the results of the second moves of "KGCPkgcp" using the normal opening rule if the first player decides to flip B-2 (cell 1). There are four possible outcomes, namely, either a king, a guard, a cannon or a pawn is revealed. Each subtable has eight columns. Column 1 shows the cell ID which is chosen by the second player, columns 2–8 list the corresponding expected difference of each remaining dark piece. Under the mini-max assumption, the second player will reveal C-2 (cell 2) as his move if the first player reveals a cannon; otherwise, the second player will reveal D-2 (cell 3).

4   Discussion

Below we discuss the benefits of the indexing function (Sect. 4.1) and the unfairness of the experimental results (Sect. 4.2).

4.1   Implementation Issues

In Sect. 2, we discussed the indexing function used in the normal version of the game. An additional benefit of the indexing function is that it reduces the size

Table 3. Experiment results of normal version of the game: the results of different first moves using different opening rules.

Index	Pieces combination	Color assigned				Normal opening			
		Block ID: 0		Block ID: 1		Block ID: 0		Block ID: 1	
		$E[x]$	σ	$E[x]$	σ	$E[x]$	σ	$E[x]$	σ
1	CCPP v.s. ccpp	−0.038	0.843	−0.026	0.844	−0.061	0.847	−0.044	0.843
2	CPPP v.s. cppp	−0.037	0.857	−0.046	0.869	−0.125	0.846	−0.030	0.871
3	GCCP v.s. gccp	−0.047	0.945	−0.053	0.948	−0.106	0.941	−0.042	0.951
4	GCPP v.s. gcpp	−0.036	0.938	−0.033	0.941	−0.075	0.933	0.010	0.942
5	GGCC v.s. ggcc	−0.058	0.958	−0.090	0.955	−0.177	0.944	−0.091	0.951
6	GGCP v.s. ggcp	−0.024	0.949	−0.043	0.946	−0.123	0.941	−0.034	0.946
7	GGMC v.s. ggmc	−0.041	0.933	−0.055	0.934	−0.136	0.924	−0.030	0.938
8	GGMM v.s. ggmm	−0.154	0.917	−0.160	0.914	−0.210	0.908	−0.137	0.909
9	GGMR v.s. ggmr	−0.108	0.912	−0.108	0.916	−0.169	0.908	−0.088	0.915
10	GMCP v.s. gmcp	−0.035	0.915	−0.043	0.920	−0.076	0.913	−0.015	0.920
11	GPPP v.s. gppp	−0.187	0.892	−0.204	0.892	−0.195	0.890	−0.177	0.894
12	KCCP v.s. kccp	−0.033	0.892	−0.044	0.890	−0.128	0.887	−0.044	0.891
13	KCPP v.s. kcpp	−0.027	0.885	−0.040	0.884	−0.121	0.877	−0.012	0.885
14	KGCC v.s. kgcc	−0.060	0.926	−0.092	0.929	−0.124	0.922	−0.044	0.932
15	KGCP v.s. kgcp	−0.041	0.924	−0.041	0.923	−0.120	0.918	0.002	0.926
16	KGGC v.s. kggc	−0.042	0.916	−0.043	0.919	−0.085	0.914	0.002	0.919
17	KGGM v.s. kggm	−0.095	0.889	−0.103	0.885	−0.103	0.890	−0.078	0.886
18	KGGP v.s. kggp	−0.156	0.924	−0.176	0.921	−0.192	0.919	−0.145	0.925
19	KGMC v.s. kgmc	−0.054	0.896	−0.047	0.904	−0.090	0.893	−0.006	0.899
20	KGMM v.s. kgmm	−0.086	0.873	−0.096	0.875	−0.096	0.876	−0.060	0.878
21	KGMP v.s. kgmp	−0.099	0.923	−0.115	0.921	−0.134	0.920	−0.084	0.922
22	KGMR v.s. kgmr	−0.067	0.876	−0.069	0.876	−0.075	0.879	−0.040	0.879
23	KPPP v.s. kppp	−0.221	0.938	−0.219	0.935	−0.288	0.927	−0.225	0.933
24	PPPP v.s. pppp	−0.486	0.874	−0.486	0.874	−0.543	0.840	−0.543	0.840

Table 4. The second move choices and results of "KGCPkgcp" after the first move in cell 1.

(a) First move flips K

	G	C	P	k	g	c	p
0	0.069	0.106	0.171	0.693	0.708	0.456	−0.642
2	0.407	−0.018	0.365	0.826	0.790	0.628	−0.594
3	0.158	0.106	0.078	0.140	0.008	0.488	0.400
4	−0.093	0.275	0.036	0.204	0.215	0.086	0.268
5	0.299	0.415	0.286	0.767	0.725	0.518	−0.636
6	0.096	0.318	0.085	−0.022	−0.146	0.114	0.136
7	0.132	0.024	0.063	0.260	−0.033	0.110	0.419

(b) First move flips G

	K	C	P	k	g	c	p
0	0.171	−0.008	0.035	−0.635	0.665	0.457	0.601
2	0.407	0.108	0.303	−0.689	0.783	0.546	0.706
3	0.065	0.150	0.079	0.293	0.110	−0.454	0.058
4	−0.090	0.256	−0.033	0.210	0.186	0.056	0.206
5	0.299	0.340	0.215	−0.635	0.731	0.478	0.618
6	0.096	0.301	0.089	0.104	−0.024	0.092	−0.139
7	0.058	0.024	0.007	0.388	0.235	0.103	−0.019

(c) First move flips C

	K	G	P	k	g	c	p
0	0.361	0.322	0.206	−0.436	−0.411	0.075	0.282
2	−0.018	−0.108	−0.083	−0.589	−0.503	0.422	−0.071
3	−0.154	−0.138	−0.201	0.754	0.765	0.497	0.657
4	0.133	0.163	0.139	0.210	0.219	0.004	0.167
5	0.415	0.340	0.313	−0.463	−0.450	0.174	0.086
6	0.318	0.301	0.333	0.015	0.050	0.003	−0.033
7	−0.192	−0.156	−0.160	0.193	0.182	0.150	0.207

(d) First move flips P

	K	G	C	k	g	c	p
0	0.075	0.131	0.044	0.725	−0.563	0.142	0.576
2	0.365	0.303	−0.083	0.722	−0.592	0.199	0.721
3	0.099	0.049	0.051	−0.017	0.293	−0.324	0.139
4	−0.058	−0.008	0.272	0.183	0.138	0.114	0.174
5	0.286	0.215	0.313	0.728	−0.544	0.272	0.663
6	0.085	0.089	0.333	−0.188	0.114	0.174	−0.001
7	0.029	0.014	0.024	−0.075	0.372	0.160	0.269

of the saved positions. In 2×4 CDC, there are at most 8 pieces. The sizes of the saved board positions with eight pieces are $2^{32} = 4,294,967,296$ and $1,441,729$ respectively for the two versions described, where the size of the latter one is only 0.03 % of the former one. The indexing function can also be used to the original CDC game.

To ensure the result of each position being correct, each computed value is verified. There are two common verifying methods. First, according to the properties of the search procedure, each position's value is returned from some of its children. Thus for each position, there should be one child position causing the value to its parent position to be recorded as it is now. Second, since the game board is symmetric and we choose pieces of the same type for both sides, a position should have the same value as the position after up-down flip, left-right flip, clock-wise rotation, or color swapping.

4.2   Experiments

In total, there are 131 possible combinations of 4 pieces. However, in the experiment, we only consider 24 of them. Although the total number of pieces combinations is 131, some pieces combinations have the same capturing relationship in between pieces. For example, "KGMP" and "KGRP" have the same relationship since the roles of K, G and P are the same in both combinations. That is, M is less than K and G but greater than P in the first combination and R is less than K and G but greater than P in the second combination. Hence, if we treat M in the first combination as R in the second combination, the two combinations can be considered as the same or isomorphic [3]. There are total 24 combinations consisting of 4 pieces that are not isomorphic.

In the oracle version of the game, we observe that the initial placement of pieces is an important factor that affects the outcome of a game. We compare the result of using the color assigned rule with the result of using the normal opening rule. We find that, in some cases, it is better to flip the opponent's piece at the beginning of the game. This is because the game is deterministic, i.e., does not involve any stochastic behaviors.

In the normal version of the game, using the color assigned rule is not always better than one using the normal opening rule for the first player, although the first revealed piece is more likely to be captured. For example, in the "KGGmkggm" combination, the best move is to reveal a king at B-2. Using the color assigned rule, the probability of revealing a king is 1/8; however, using the normal opening rule, the probability of revealing a king increases to 1/4.

Although the experiment results show the unfairness of the game due to the position where the first move is made, under some conditions, the initial placement of the pieces lead to a fair result. That is, both sides have an equal chance of winning. Figure 4 shows the three fairest board positions with two revealed pieces and restricted the next move in the "KGMPkgmp" combination.

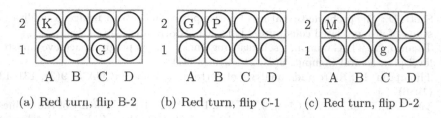

(a) Red turn, flip B-2   (b) Red turn, flip C-1   (c) Red turn, flip D-2

Fig. 4. Fair board positions after a sequence of flips in the normal version "KGMP-kgmp" game.

5   Conclusion

In this paper, we studied CDC on a 2 × 4 game board. We have solved the oracle version and the normal version of the game with two different opening rules, the color assigned rule and the normal opening rule. We have also exploited fundamental properties of CDC. In the implementation stage, we use an highly improved indexing function to avoid the ambiguity and save 99.97 % of the storage space. In particular, we solved the 2 × 4 CDC for the 24 fair combinations of 4 pieces, where both players have identical pieces, i.e., 4 pieces each. The experimental results show that the initial positions of the pieces and the choice of where to make the first move are the key factors that affect a game's outcome. In the oracle version, most games favor the first player. In the normal version, most games favor the second player. However, in some situations, under the normal opening rule, the first player has a higher probability of revealing a piece that is good for the first player. This is because the probability of getting a good flip under the normal opening rule is twice of that when the color assigned rule is used. The results also show that although the beginning of the game is not fair, in some specified openings, it is fair.

References

1. Chen, B.-N., Shen, B.-J., Hsu, T.-S.: Chinese dark chess. ICGA J. **33**(2), 93–106 (2010)
2. Chen, J.-C., Lin, T.-Y., Hsu, S.-C., Hsu, T.-s.: Design and implementation of computer chinese dark chess endgame database. In: Proceedings of TCGA Workshop, pp. 5–9 (2012)
3. Chen, J.-C., Lin, T.-Y., Hsu, T.-s.: Equivalence partition of dark chess endgames (submitted) (2013)
4. Knuth, D.E.: The art of computer programming. In: Fascicle 0: Introduction to Combinatorial Algorithms and Boolean Functions, vol. 4, 1st edn. Addison-Wesley Professional, New York (2008)
5. Korf, R.E.: Depth-first iterative-deepening: an optimal admissible tree search. Artif. Intell. **27**(1), 97–109 (1985)
6. Lin, C.-H., Shan, Y.-C., Wu, I-C.: Tournament framework for computer mahjong competitions. In: 2011 International Conference on Technologies and Applications of Artificial Intelligence (TAAI), pp. 286–291. IEEE (2011)

7. Scott, D.: The problematic nature of participation in contract bridge: a qualitative study of group-related constraints. Leis. Sci. **13**(4), 321–336 (1991)
8. Tesauro, G.: TD-Gammon, a self-teaching backgammon program, achieves master-level play. Neural Comput. **6**(2), 215–219 (1994)
9. Thompson, K.: Retrograde analysis of certain endgames. ICCA J. **9**(3), 131–139 (1986)
10. Yen, S.-J., Chou, C.-W., Chen, J.-C., Wu, I.-C., Kao, K.-Y.: The art of the Chinese dark chess program diable. In: Chang, R.-S., Jain, L.C., Peng, S.-L. (eds.) Advances in Intelligent Systems and Applications - Volume 1. Smart Innovation, Systems and Technologies, vol. 20, pp. 231–242. Springer, Berlin Heidelberg (2013)

Cylinder-Infinite-Connect-Four Except for Widths 2, 6, and 11 Is Solved: Draw

Yoshiaki Yamaguchi[✉], Tetsuro Tanaka, and Kazunori Yamaguchi

The University of Tokyo, Tokyo, Japan
yoshiaki@graco.c.u-tokyo.ac.jp

Abstract. Cylinder-Infinite-Connect-Four is a variant of the Connect-Four game played on cylindrical boards of differing cyclic widths and infinite height. In this paper, we show strategies to avoid losing at Cylinder-Infinite-Connect-Four except for Widths 2, 6, and 11. If both players use the strategies, the game will be drawn. This result can also be used to show that Width-Limited-Infinite-Connect-Four is drawn for any width. We also show that Connect-Four of any size with passes allowed is drawn.

1 Introduction

Solving games has long been one of the main targets of game research [9]. Infinite-Connect-Four is a variant of the Connect-Four game played on an infinite size board. Cylinder-Infinite-Connect-Four is another variant of Connect-Four played on cylindrical boards. First, we introduce a definition of Cylinder-Infinite-Connect-Four.

Definition 1. Cylinder-Infinite-Connect-Four: *Connect-Four with a horizontally rotating board of infinite height and finite width.*

In this paper, we introduce a newly obtained solution for Cylinder-Infinite-Connect-Four in the form of cannot-lose strategies for both players. The players can use the strategies to prevent the opponent from achieving a won game. This means that optimal play by both players will result in a draw.

Next, we introduce the rules of Cylinder-Infinite-Connect-Four.

Cylinder-Infinite-Connect-Four is a game for two players, with the first player designated as X and the second player as O. The game is played on a cylindrical board consisting of cells piled up infinitely in height and cyclically in width. Cells are arranged in horizontal rows cyclically numbered beginning with zero or in vertical columns numbered beginning with one. Each player places disks marked with their symbol (X or O) in the cells. Disks gravitate and so must be placed in the lowest unoccupied cell for each column. The notation "X moves *i*" means that X places a disk in column *i*. The object of the game is to connect at least four of one's own disks next to each other horizontally, vertically, or diagonally before the opponent. Such a connection is called a *Connect 4*. One disk is counted only once for each Connect 4. A player who achieves a Connect 4

H.J. van den Herik et al. (Eds.): CG 2013, LNCS 8427, pp. 163–174, 2014.
DOI: 10.1007/978-3-319-09165-5_14, © Springer International Publishing Switzerland 2014

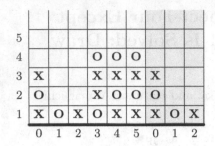

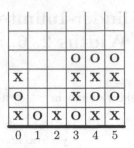

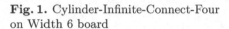

Fig. 1. Cylinder-Infinite-Connect-Four on Width 6 board

Fig. 2. Infinite-Connect-Four on Width 6 board

wins the game. If it becomes impossible for either X or O to achieve a Connect 4 and the game never ends, the game is *drawn*.

We call a configuration of disks a *position*. A disk already in a position is denoted in boldface and a disk to be placed in a position is denoted in italics.

An example position in Cylinder-Infinite-Connect-Four for Width 6 is shown in Fig. 1. A Connect 4 in columns 3-0, 4-1, or 5-2 is not visually connected. To make it easy to see such a Connect 4, we duplicate columns 0-2 to make the Connect 4 visually connected. The duplicated columns are shown in gray in Fig. 1. If there is a Connect 4 of X or O in a board of Cylinder-Infinite-Connect-Four, the Connect 4 is included in some 4 × 4 sub-board of the board. So, if there is no Connect 4 of X or O in all the 4 × 4 subboards in the board, there is no Connect 4 of X or O in the board.

In Cylinder-Infinite-Connect-Four players have more opportunities to achieve a Connect 4 than in Infinite-Connect-Four with the same width because in the former a Connect 4 may wrap around the rightmost column to the leftmost column. For example, in the Cylinder-Infinite-Connect-Four position in Fig. 2, which is the same as the Cylinder-Infinite-Connect-Four position in Fig. 1, X did not achieve a Connect 4.

There are local rules in Infinite-Connect-Four forbidding players to place a disk far from the already placed disks [14]. In this paper, we use the "Free Placement Rule" that places no restriction on disk placement.

On the cylindrical board for Width 1, X's first disk is adjacent to oneself but a single disk does not make a Connect 4 by the rule of counting one disk only once for each Connect 4. So, this game is a draw.

2   Previous Work

In 1998, it was proved by Allen [1] that in Connect-Four played on the standard board of Width 7 and Height 6 X wins; independently and almost simultaneously it was also proven, too, by Allis [2]. The results of the games played on finite boards with non-standard heights and widths were reported in [1,12]. We proved that in Connect-Four played on a board infinite in height, width, or both, optimal play by both players, i.e., both players use cannot-lose strategies, leads to a draw [13]. These cannot-lose strategies are based on paving similar to that used in polyomino achievement games [4,6–8] and 8 (or more) in a row [15]. There are a number of cyclic games including PAC-MAN [10], Cylinder Go [5], Torus Go [5], Torus and Klein bottle games [11], and n^d Torus Tic-Tac-Toe [3], but as far as we know the properties of cyclic games have not been studied much yet.

3   Solution for Cylinder-Infinite-Connect-Four Except for Widths 2, 6, and 11

In this section, we show cannot-lose strategies for playing Cylinder-Infinite-Connect-Four for both players. First, we introduce definitions of terms used in stating the results.

Definition 2. Tile: *a pair of cells used to block cells to be occupied by opponent's disks.*

We use four kinds of tiles: adjacent-tiles, space-tiles, vertical-tiles, and sky-tiles. Next, we define adjacent-tiles and space-tiles.

Definition 3. Adjacent-tile: *a tile placed horizontally and composed of two adjacent cells. Both cells of an adjacent-tile are denoted by A or B.*

The adjacent-tile in Fig. 3 is bordered in boldface.

Definition 4. Space-tile: *a tile placed horizontally and composed of two cells separated by another cell between them. Both cells of a space-tile are denoted by S or T.*

The space-tile in Fig. 4 is bordered in boldface.

Next, we define terms describing methods of play, the two other types of tiles, and other terms.

Definition 5. Paving the tiles: *forcing one's opponent to place at most one disk in each tile of a group of tiles.*

Definition 6. Follow-in-tile: *the act of placing one's disk in a tile after the opponent has placed a disk in the tile.*

An adjacent-tile in the first row can be paved by playing a follow-in-tile. In Fig. 5, O played a follow-in-tile in an adjacent-tile.

Fig. 3. Adjacent-tile.

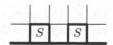

Fig. 4. Space-tile.

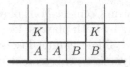

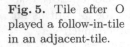

Fig. 5. Tile after O played a follow-in-tile in an adjacent-tile.

Fig. 6. Vertical-tile.

Fig. 7. CUP

Definition 7. Vertical-tile: *a tile placed vertically and composed of two adjacent cells. A vertical-tile is denoted by two disks placed so that the player's disk is over the opponent's disk.*

Definition 8. Follow-up: *the act of placing one's disk in a vertical-tile after the opponent has placed a disk in it [2].*

Figure 6 shows a vertical-tile for O. When a player plays a follow-up in a vertical-tile, the player's disk is over the opponent's disk because of gravity.

Definition 9. Sky-tile: *a tile other than an adjacent-tile, a space-tile, and a vertical-tile. Both cells of a sky-tile are denoted by K.*

Definition 10. Initial-cell: *the cell for X's first move. The initial-cell is placed in the first row and zeroth column of a board.*

Definition 11. Free-cell: *a cell that may be occupied by either X or O. A free-cell is denoted by F.*

Definition 12. CUP: *a cup-shaped combination of two adjacent-tiles on the ground and one sky-tile as illustrated in Fig. 7.*

Lemma 1. *[13] A CUP can be paved.*

The following is the main theorem of this paper.

Theorem 1. *Cylinder-Infinite-Connect-Four except for Widths 2, 6, and 11 is solved: Optimal play by both players leads to a draw.*

We prove Theorem 1 by showing cannot-lose strategies for both players for playing Cylinder-Infinite-Connect-Four except for Widths 2, 6, and 11.

Next, we define CUP-5, Cylinder-X-Board, and Cylinder-O-Board to show cannot-lose strategies for Cylinder-Infinite-Connect-Four.

Definition 13. *CUP-5: a board of infinite height and width 5 with a CUP in the left four columns filled with vertical tiles in other cells.*

A CUP-5 for X is shown in Fig. 8. By exchanging X and O, we get a CUP-5 for O.

Table 1. Cylinder-X-Boards except for Widths 2 and 6

Width	1	2	3	4+5n	5
X-Board	Follow-ups	-	Figure 11	Figure 13	Figure 14
CUP-5	-	-	-	2|3	-
Width	6	7	8+5n	-	10+5n
X-Board	-	Figure 15	Figure 16	-	Figure 18
CUP-5	-	-	3|4	-	5|6
Width	11+5n	12+5n	-	-	-
X-Board	Figure 19	Figure 20	-	-	-
CUP-5	6|7	6|7	-	-	-

Definition 14. Cylinder-X-Board for Width n: *a combination of the initial-cell, tiles, free-cells, and CUPs that determines a cannot-lose strategy for X in Cylinder-Infinite-Connect-Four for Width n.*

Table 1 lists Cylinder-X-Boards except for Widths 2, 6, and 11. X plays only follow-ups above the second row in Cylinder-X-Boards. The attribute 'X-Board' has the figure number of the cannot-lose strategy for X in the Cylinder-X-Board of the width specified in the attribute 'Width'.

By inserting a CUP-5 between two columns in each Cylinder-X-Board shown as the attribute 'CUP-5' in Table 1, each Cylinder-X-Board can be increased by five columns. For example, we can make a Cylinder-X-Board for Width 9 by inserting a CUP-5 between the second and third columns of a Cylinder-X-Board for Width 4 (Fig. 17) .

There is no Connect 4 of X or O in all the Cylinder-X-Boards in Table 1.

Definition 15. *Joint: five columns consisted of three columns from a CUP-5 or a Cylinder-X(O)-Board before the CUP-5 insertion and two columns from the adjacent CUP-5. In a Joint, a sky-tile is replaced by free-cells.*

Figure 9 shows a Joint. All the 4×4 subboards by inserting CUP-5s into Cylinder-X-Board are present in Figs. 8 and 9. There is no Connect 4 of X or O in Figs. 8 and 9. So, there is no Connect 4 of X or O in all the Cylinder-X-Boards except for Widths 2 and 6.

Definition 16. Cylinder-O-Board for Width n: *a combination of tiles, free-cells, and CUPs that determines a cannot-lose strategy for O in Cylinder-Infinite-Connect-Four for Width n.*

Table 2 lists Cylinder-O-Boards except for Widths 2, 6, and 11. O plays only follow-ups above the third row in Cylinder-O-Boards. The attributes 'Width', 'O-Board', and 'CUP-5' are the same as those in Table 1. In Cylinder-O-Boards for Widths 7 and 9, O should move 1 as the first move of O.

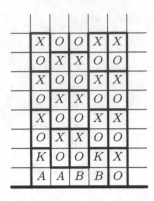

O	X	X	X	O
X	O	O	O	X
O	X	X	X	O
X	O	O	O	X
O	X	X	X	O
X	O	O	O	X
O	F	X	F	O
A	A	O	A	A

Fig. 8. CUP-5 for Cylinder-X-Board.   **Fig. 9.** Joint in a Cylinder-X-Board.

Table 2. Cylinder-O-Boards except for Widths 2, 6, and 11

Width	1	2	3	4	5+5n
O-Board	Follow-ups	Follow-ups	Figure 12	Figure 21	Figure 22
CUP-5	-	-	-	-	4|0
Width	6	7+5n	8+5n	9+5n	-
O-Board	-	Figure 23	Figure 24	Figure 26	-
CUP-5	-	6|0	0|1	6|7	-
Width	11	-	-	-	-
O-Board	-	-	-	-	-
CUP-5	-	-	-	-	-
Width	16+5n	-	-	-	-
O-Board	Figure 25	-	-	-	-
CUP-5	15|0	-	-	-	-

There is no Connect 4 of X or O in Cylinder-O-Board in Table 2. By inserting CUP-5s into Cylinder-O-Board, 4×4 subboards are present in Figs. 8 and 9 with X and O exchanged.

There is no Connect 4 of X or O in all the Cylinder-O-Boards except for Widths 6 and 11. Cylinder-X-Boards and Cylinder-O-Boards can be paved in the following ways. A CUP is paved by Lemma 1. A player plays a follow-in-tile when his opponent moves in an empty tile. When the opponent moves in a free-cell or a cell of a tile which one disk is in, the player moves in another free-cell or a cell in other tiles.

There are some special cases for Cylinder-O-Board for Width 9. If X moved in F in the fifth column and no disk is in the seventh, eighth, zeroth, and the first columns except for the initial-cell and the cell to its right, O moves in K in the first column and replaces another K by F. If X moved in F in the fifth

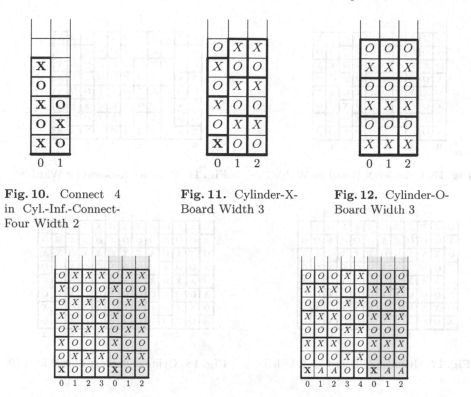

Fig. 10. Connect 4 in Cyl.-Inf.-Connect-Four Width 2

Fig. 11. Cylinder-X-Board Width 3

Fig. 12. Cylinder-O-Board Width 3

Fig. 13. Cylinder-X-Board for Width 4

Fig. 14. Cylinder-X-Board for Width 5

column and A in the eighth column is occupied by O and K in the first column is occupied by X, O moves in F in the first column.

Because the number of cells except for vertical-tiles is odd in each Cylinder-X-Board, X can force O to place a disk in a vertical-tile first.

Because the number of cells except for vertical-tiles is even in each Cylinder-O-Board, O can force X to place a disk in a vertical-tile first.

Cylinder-Infinite-Connect-Fours for Widths 2, 6, and 11 are not yet solved. Cylinder-Infinite-Connect-Four for Width 2 has only two columns but is strikingly difficult. For example, O's follow-ups can prevent X from achieving a Connect 4, but X's follow-ups cannot prevent O from achieving a Connect 4 spirally if O moves 0, 0, 1, and 1 as shown in Fig. 10.

4   Solution for Infinite-Connect-Four for Finite Width

We stated the following theorem in [13], but we omitted the proof because of space limitations.

Theorem 2. *Infinite-Connect-Four with any width is solved: Optimal play by both players leads to a draw.*

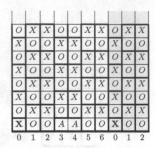

Fig. 15. Cylinder-X-Board for Width 7

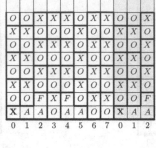

Fig. 16. Cylinder-X-Board for Width 8

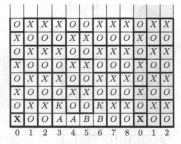

Fig. 17. Cylinder-X-Board for Width 9

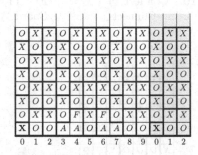

Fig. 18. Cylinder-X-Board for Width 10

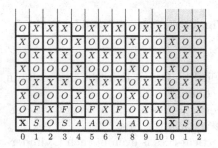

Fig. 19. Cylinder-X-Board for Width 11

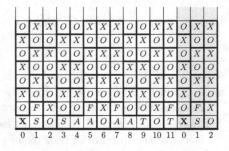

Fig. 20. Cylinder-X-Board for Width 12

Here, we prove Theorem 2 by presenting strategies for both players to play Infinite-Connect-Four using Theorem 1.

Definition 17. Width-Limited-X-Board for Width n: *the cannot-lose strategy for X in Infinite-Connect-Four for Width n.*

Definition 18. Width-Limited-O-Board for Width n: *the cannot-lose strategy for O in Infinite-Connect-Four for Width n.*

Proof. In Infinite-Connect-Four for Width 2, X and O can achieve a draw only by playing follow-ups.

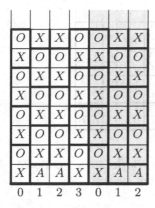

Fig. 21. Cylinder-O-Board for Width 4

O	X	X	O	O	X	X
X	O	O	X	X	O	O
O	X	X	O	O	X	X
X	O	O	X	X	O	O
O	X	X	O	O	X	X
X	O	O	X	X	O	O
O	X	X	O	O	X	X
X	A	A	X	X	A	A

0 1 2 3 0 1 2

Fig. 22. Cylinder-O-Board for Width 5

O	X	X	O	O	O	X	X
X	O	O	X	X	X	O	O
O	X	X	O	O	O	X	X
X	O	O	X	X	X	O	O
O	X	X	O	O	O	X	X
X	O	O	X	X	X	O	O
K	X	X	K	O	K	X	X
A	A	B	B	X	A	A	B

0 1 2 3 4 0 1 2

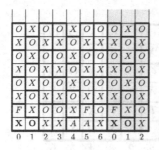

Fig. 23. Cylinder-O-Board for Width 7

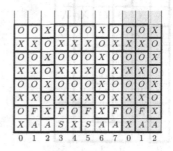

Fig. 24. Cylinder-O-Board for Width 8

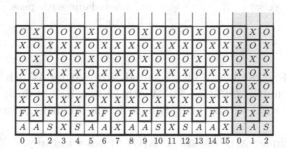

Fig. 25. Cylinder-O-Board for Width 16

In Infinite-Connect-Four for width more than 3 except for 6 and 11, X and O can achieve a draw by moving along their Cylinder-X(O)-Board with the specified width.

In Infinite-Connect-Four for Width 6, X can achieve a draw by moving along the Width-Limited-X-Board for Width 6 in Fig. 27. In Infinite-Connect-Four for Width 6, O can achieve a draw by moving along the Width-Limited-O-Board

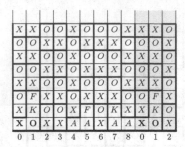

Fig. 26. Cylinder-O-Board for Width 9

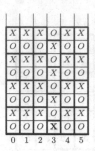

Fig. 27. Width-Limited-O-Board for Width 6 (not cyclic)

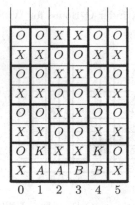

Fig. 28. Width-Limited-O-Board for Width 6 (not cyclic)

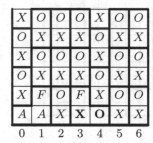

Fig. 29. O's cannot-lose strategy in Connect-Four with pass when X moves 3 initially

for Width 6 in Fig. 28. The Width-Limited-X-Board for Width 11 is shown in Fig. 19. Width-Limited-O-Board for Width 11 can be obtained by inserting a CUP-5 into the right of the fifth column in the Width-Limited-O-Board for Width 6.

For each Cylinder-X(O)-Board, if X(O) moves i initially, a Width-Limited-X(O)-Board is constructed by rotating columns cyclically so that the board's initial-cell is in column i. For example, in Infinite-Connect-Four with Width 9, if X moves 8 initially, the columns in the Cylinder-O-Board are rotated by one column left so that the initial-cell **X** is in the eighth column and the zeroth to eighth columns are used as the Width-Limited-O-Board.   ∎

Now we turn to Connect-Four with passes allowed.

Definition 19. *'Connect-Four with passing': Connect-Four played such that either player can pass in turn instead of placing a disk.*

If the players pass consecutively, they should pass again on the next turn because the position does not change from the previous turn. This leads to a draw.

From Theorem 2, we obtain the following corollary.

Corollary 1. *Connect-Four in any height and width with passing is solved: Optimal play by both players leads to a draw.*

Proof. We show cannot-lose strategies for X and O. In the following all strategies, if an opponent declares a pass, a player declares a pass again and the game is drawn.

In a Width-Limited-O-Board for Height 1 (2), adjacent-tiles are placed in the first row (and second row) from the leftmost column and free-cells are placed in the rightmost column if the width of the board is odd. The cannot-lose strategy is as follows. If X places a disk in a cell in an adjacent-tile, O plays a follow-in-tile in the adjacent-tile. If X places a disk in a free-cell, O declares a pass.

Width-Limited-X-Board for Height 1 (2) is the same as Width-Limited-O-Board for Height 1 (2) except that X declares a pass initially.

A Width-Limited-X-Board for Height 3 or more is made from the Cylinder-X-Board for the same width truncated to the specified height h. In the truncated Cylinder-X-Board, vertical-tiles spanning the rows h and $h + 1$ are chopped. If O moves in one of the chopped vertical-tiles, X declares a pass.

A Width-Limited-O-Board for Height 3 or more is made similarly.   ∎

For example, in the standard board, if X moves 3 and O moves 4, O's cannot-lose strategy is as shown in Fig. 29.

5   Conclusion

We proved that Cylinder-Infinite-Connect-Four except for Widths 2, 6, and 11 is drawn. From this, we proved that Infinite-Connect-Four with limited width is drawn. We also proved that Connect-Four in any height and width with passes allowed is drawn.

Our future work is to solve Cylinder-Infinite-Connect-Four for Widths 2, 6, and 11 to complete the solution.

Acknowledgement. We are grateful to the reviewers for their valuable suggestions regarding Cylinder-Infinite-Connect-Four and passing.

References

1. Allen, J.D.: A note on the computer solution of Connect-Four. In: Levy, D.N.L., Beal, D.F. (eds.) Heuristic Programming in Artificial Intelligence, The First Computer Olympiad, pp. 134–135. Ellis Horwood, Chinchester (1989)
2. Allis, L.V.: A knowledge-based approach to Connect-Four. The game is solved: white wins. Master's thesis, Vrije Universiteit (1988)

3. Beck, J.: Positional games. Comb. Probab. Comput. **14**, 649–696 (2005)
4. Gardner, M.: Mathematical games. Sci. Am. **240**, 18–26 (1979)
5. Geselowitz, L.: Freed Go. http://www.leweyg.com/lc/freedgo.html
6. Halupczok, I., Puchta, J.C.S.: Achieving snaky. Integers: Electron. J. Comb. Number Theor. **7**, G02 (2007)
7. Harary, F.: Is snaky a winner? Geombinatorics **2**, 79–82 (1993)
8. Harary, F., Harborth, H.: Achievement and avoidance games with triangular animals. J. Recreat. Math. **18**(2), 110–115 (1985-1986)
9. van den Herik, H.J., Uiterwijk, J.W.H.M., van Rijswijck, J.: Games solved: now and in the future. Artif. Intell. **134**(1–2), 277–311 (2002)
10. NAMCO, PAC-MAN. http://pacman.com
11. Torus and Klein Bottle Games. http://www.math.ntnu.no/dundas/75060/TorusGames/TorusGames.html
12. Tromp, J.: Solving Connect-4 on medium board sizes. ICGA J. **31**(2), 110–112 (2008)
13. Yamaguchi, Y., Yamaguchi, K., Tanaka, T., Kaneko, T.: Infinite Connect-Four is solved: draw. In: van den Herik, H.J., Plaat, A. (eds.) ACG 2011. LNCS, vol. 7168, pp. 208–219. Springer, Heidelberg (2012)
14. Yasda, H.: Hakotumi (Infinite-Connect-Four with Gravity). http://homepage3.nifty.com/yasda/dwnload/hako.htm
15. Zetters, T.G.L.: 8 (or more) in a row. Am. Math. Monthly **87**, 575–576 (1980)

Havannah and TwixT are PSPACE-complete

Édouard Bonnet, Florian Jamain, and Abdallah Saffidine[✉]

LAMSADE, Université Paris-Dauphine, Paris, France
{edouard.bonnet,abdallah.saffidine}@dauphine.fr,
florian.jamain@lamsade.dauphine.fr

Abstract. Numerous popular abstract strategy games ranging from HEX and HAVANNAH via TWIXT and SLITHER to LINES OF ACTION belong to the class of connection games. Still, very few complexity results on such games have been obtained since HEX was proved PSPACE-complete in the early 1980s.

We study the complexity of two connection games among the most widely played ones, i.e., we prove that HAVANNAH and TWIXT are PSPACE-complete. The proof for HAVANNAH involves a reduction from GENERALIZED GEOGRAPHY and is based solely on ring-threats to represent the input graph. The reduction for TWIXT builds up on previous work as it is a straightforward encoding of HEX.

1 Introduction

A connection game is a kind of abstract strategy game in which players try to make a specific type of connection with their pieces [5]. In many connection games, the goal is to connect two opposite sides of a board. In these games, players take turns placing or/and moving pieces until they connect the two sides of the board. HEX, TWIXT, and SLITHER are typical examples of this type of game. However, a connection game can also involve completing a loop (HAVANNAH) or connecting all the pieces of a color (LINES OF ACTION).

A typical process in studying an abstract strategy game, and in particular a connection game, is to develop an artificial player for it by adapting standard techniques from the game-search literature, in particular the classical Alpha-Beta algorithm [1] or the more recent Monte-Carlo Tree Search paradigm [2,6]. These algorithms explore an exponentially large game tree and are meaningful when optimal polynomial time algorithms are impossible or unlikely. For instance, tree-search algorithms would not be used for NIM and SHANNON'S EDGE SWITCHING GAME which can be played optimally and solved in polynomial time [7].

The complexity class PSPACE comprises those problems that can be solved on a Turing machine using an amount of space polynomial in the size of the input. The prototypical example of a PSPACE-complete problem is the Quantified Boolean Formula problem (QBF) which can be seen as a generalization of SAT allowing for variables to be both existentially and universally quantified. Proving that a game is PSPACE-hard shows that a variety of intricate problems can be

H.J. van den Herik et al. (Eds.): CG 2013, LNCS 8427, pp. 175–186, 2014.
DOI: 10.1007/978-3-319-09165-5_15, © Springer International Publishing Switzerland 2014

encoded via positions of this game. Additionally, it is widely believed in complexity theory that if a problem is PSPACE-hard, then it admits no polynomial time algorithms.

For this reason, studying the computational complexity of games is a popular research topic. The complexity class of CHESS and GO was determined shortly after the very definition of these classes and other popular games have been classified since then [8,12]. More recently, we studied the complexity of trick taking card games which notably include BRIDGE, SKAT, TAROT, and WHIST [4].

Connection games have received less attention. Besides Even and Tarjan's proof that SHANNON'S VERTEX SWITCHING GAME is PSPACE-complete [9] and Reisch's proof that HEX is PSPACE-complete [20], the only complexity results on connection games that we know of are the PSPACE-completeness of virtual connection detection [14] in HEX, the NP-completeness of dominated cell detection in SHANNON'S VERTEX SWITCHING GAME [3], as well as an unpublished note showing that a problem related to TWIXT is NP-complete [18].[1]

The two games that we study in this paper (HAVANNAH and TWIXT) rank among the most notable connection games. They were the main topic of multiple master's theses and research articles [10,13,16,18,19,22,23], and they both gave rise to competitive play. High-level online competitive play takes place on www.littlegolem.net. Finally, live competitive play can also be observed between human players at the Mind Sports Olympiads where an international TWIXT championship has been organized every year since 1997, as well as between HAVANNAH computer players at the ICGA Computer Olympiad since 2009.[2]

2   HAVANNAH

HAVANNAH is a 2-player connection game played on a hexagonal board paved by hexagons. White and Black place a stone of their color in turn in an unoccupied cell. Stones can neither be taken, nor moved, nor removed. Two cells are neighbors if they share an edge. A group is a connected component of stones of the same color via the neighbor relation. A player wins if they realize one of the three following different structures: a circular group, called *ring*, with at least one cell, possibly empty, inside; a group linking two corners of the board, called *bridge*; or a group linking three edges of the board, called *fork*.

As the length of a game of HAVANNAH is polynomially bounded, exploring the whole game tree can be done with polynomial space, so HAVANNAH is in PSPACE.

In our reduction, the HAVANNAH board is sufficiently large so that the gadgets are far from the edges and the corners. Additionally, the gadgets feature ring threats that are sufficiently short so that the bridges and forks winning conditions do not have any influence. Before starting the reduction, we define threats and make two observations that will prove useful in the course of the reduction.

---

[1] For a summary in English of Reisch's reduction, see Maarup's thesis [17].

[2] See www.boardability.com/game.php?id=twixt and www.grappa.univ-lille3.fr/icga/game.php?id=37 for details.

A *simple threat* is defined as a move which threatens to realize a ring on the next move on a unique cell. There are only two kinds of answers to a simple threat: either win on the spot or defend by placing a stone in the cell creating this very threat. A *double threat* is defined as a move which threatens to realize a ring on the next move on at least two different cells. We will use *threat* as a generic term to encompass both simple and double threats. A *winning sequence of threats* is defined as a sequence of simple threats ended by a double threat for one player such that the opponent's forced move never makes a threat. Thus, when a player is not threatened and can initiate a winning sequence of threats, they do win. To be more concise, we will denote by $W : a_1,a_2; a_3,a_4; \ldots; a_{2n-1}(,a_{2n})$ the sequence of moves starting with White's move a_1, Black's answer a_2, and so on. a_{2n} is optional, for the last move of the sequence might be White's or Black's. Similarly, $B : a_1,a_2; a_3,a_4; \ldots; a_{2n-1}(,a_{2n})$ denotes the corresponding sequence of moves initiated by Black. We will use the following lemmas multiple times.

Lemma 1. *If a player is not threatened, playing a simple threat forces the opponent to answer on the cell of the threat.*

Lemma 2. *If a player is not threatened, playing a double threat is winning.*

2.1 GENERALIZED GEOGRAPHY

GENERALIZED GEOGRAPHY (GG) is one of the first two-player games to have been proved PSPACE-complete [21]. It has been used to reduce to multiple games including HEX, OTHELLO, and AMAZONS [8,11,20]. Below we use GG to reduce it to HAVANNAH and to prove that HAVANNAH is PSPACE-hard (Theorem 1).

In GG, players take turns moving a token from vertex to vertex. If the token is on a vertex v, then it can be moved to a vertex v' neighboring v provided v' has not been visited yet. A player wins when it is their opponent's turn and the opponent has no legal moves. An instance of GG is a graph G and an initial vertex v_0, and asks whether the first player has a winning strategy in the corresponding game.

We denote by $P(v)$ the set of predecessors of the vertex v in G, and $S(v)$ the set of successors of v. A vertex with in-degree i and out-degree o is called (i, o)-vertex. The degree of a vertex is the sum of the in-degree and the out-degree, and the degree of G is the maximal degree among all vertices of G. If V is the set of vertices of G and V' is a subset of vertices, then $G[V \setminus V']$ is the induced subgraph of G where vertices belonging to V' have been removed.

Lichtenstein and Sipser have proved that the game remained PSPACE-hard even if G was assumed to be bipartite and of degree at most 3 [15]. We will reduce from such a restriction of GG to show that HAVANNAH is PSPACE-hard. To limit the number of gadgets we need to create, we will also assume a few simplifications detailed below. An example of a simplified instance of GG can be found in Fig. 1.

Let (G, v_0) be an instance of GG with G bipartite and of degree at most 3. We can assume that there is no vertex v with out-degree 0 in G. Indeed, if $v_0 \in P(v)$

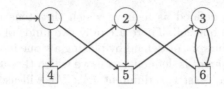

Fig. 1. Example of an instance of GG with vertex 1 as initial vertex.

then (G, v_0) is trivially winning for Player 1. Else, $(G[V \setminus (\{v\} \cup P(v))], v_0)$ is an equivalent instance, since playing in a predecessor of v is losing.

All edges coming to the initial vertex v_0 can be removed to form an equivalent instance. So, v_0 is a $(0,1)$-, a $(0,2)$-, or a $(0,3)$-vertex. If $S(v_0) = \{v'\}$, then $(G[V \setminus \{v_0\}], v')$ is a strictly smaller instance such that Player 1 is winning in (G, v_0) if and only if Player 1 is losing in $(G[V \setminus \{v_0\}], v')$. If $S(v_0) = \{v', v'', v'''\}$, then Player 1 is winning in (G, v_0) if and only if Player 1 is losing in at least one of the three instances $(G[V \setminus \{v_0\}], v')$, $(G[V \setminus \{v_0\}], v'')$, and $(G[V \setminus \{v_0\}], v''')$. In those three instances v', v'', and v'' are not $(0,3)$-vertices since they had in-degree at least 1 in G. Therefore, we can also assume that v_0 is $(0,2)$-vertex.

We call an instance with an initial $(0,2)$-vertex and then only $(1,1)$-, $(1,2)$-, and $(2,1)$-vertices a *simplified* instance.

In the following subsections we propose gadgets that encode the different parts of a *simplified* instance of GG. These gadgets have starting points and ending points. The gadgets are assembled so that the ending point of a gadget coincides with the starting point of the next one. The resulting instance of HAVANNAH is such that both players must enter in the gadgets by a starting point and leave it by an ending point otherwise they lose.

2.2 Edge Gadgets

Wires, curves, and crossroads will enable us to encode the edges of the input graph. In the representation of the gadgets, White and Black stones form the proper gadget. Dashed stones and gray stones are respectively White and Black stones setting the context.

In the HAVANNAH board we name the 6 directions: North, North-West, South-West, South, South-East, and North-East according to standard designation. While figures and lemmas are mostly presented from White's point of view, all the gadgets and lemmas work exactly the same way with colors reversed.

The wire gadget. Basically, a wire teleports moves: one player plays in a cell u and their opponent has to answer in a possibly remote cell v. u is called the starting point of the wire and v is called its ending point. A wire where White prepares a threat and Black answers is called a WB-wire (Fig. 2a); conversely, we also have BW-wires. We say that WB-wires and BW-wires are *opposite* wires. Note that wires can be of arbitrary length and can be curved with 120° angles

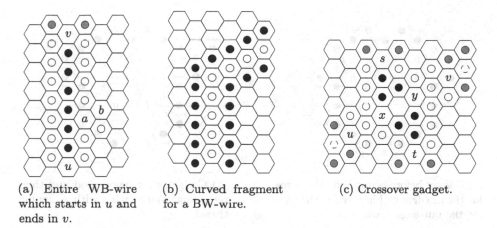

(a) Entire WB-wire which starts in u and ends in v.

(b) Curved fragment for a BW-wire.

(c) Crossover gadget.

Fig. 2. Edge gadgets.

(Fig. 2b). On an empty board, a wire can link any pair of cells as starting and ending point provided they are sufficiently far from each other.

Lemma 3. *If White plays in the starting point u of a WB-wire (Fig. 2a), and Black does not answer by a threat, Black is forced to play in the ending point v (possibly with moves at a and b interleaved).*

The crossover gadget. The input graph of GG might not be planar, so we have to design a crossover gadget to enable two chains of wires to cross. Figure 2c displays a crossover gadget, we have a South-West BW-wire with starting point u which is linked to a North-East BW-wire with ending point v, and a North BW-wire with starting point s is linked to a South BW-wire with ending point t.

Lemma 4. *In a crossover gadget (Fig. 2c), if White plays in the starting point u, Black ends up playing in the ending point v and if White plays in the starting point s, Black ends up playing in the ending point t.*

Proof. By Lemma 1, if White plays in u, Black has to play in x, forcing White to play in y, forcing finally Black to play in v. If White plays in s, again by Lemma 1, Black has to play in t. □

Note that the South wire is linked to the North wire irrespective of whether the other pair of wires has been used and conversely. That is, in a crossover gadget two paths are completely independent.

2.3 Vertex Gadgets

We now describe the gadgets encoding the vertices. Recall from Sect. 2.1 that simplified GG instances only feature $(1,2)$-, $(1,1)$-, and $(2,1)$-vertices, and a $(0,2)$-vertex. One can encode a $(1,1)$-vertex with two consecutive opposite wires.

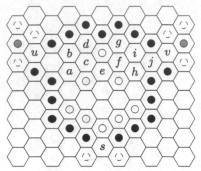

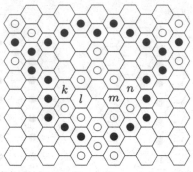

(a) Gadget before being used. The wires for the in-edges end in u and v, the wire for the out-edge starts in s.

(b) Gadget after being used and then reentered. White wins with a double threat.

Fig. 3. The (2,1)-vertex gadget links three WB-wires. The North-West and North-East ones end in u and v, the South WB-wire starts in s.

Thus, we will only present three vertex gadgets, one for $(2, 1)$-vertices, one for $(1, 2)$-vertices, and one for the $(0, 2)$-vertex.

The (2,1)-vertex gadget. A $(2, 1)$-vertex gadget receives two wire ending points. If a stone is played on either of those ending points, it should force an answer in the starting point of a third wire. That simulates a vertex with two edges going in and one edge going out.

Lemma 5. *If Black plays in one of the two possible starting points u and v of a (2,1)-vertex gadget (Fig. 3b), and White does not answer by a threat, White is forced to play in the ending point s.*

Proof. Assume Black plays in u and White answers by a move which is neither in s nor a threat. This move from White has to be either in v or in j, otherwise, Black has a double threat by playing in s and wins by Lemma 2. Now assume White plays in v. Now, Black plays in s with a simple threat in j, so White has to play in j by Lemma 1. Then Black has the following winning sequence: B: a,b; c,d; h,i; f. Black has now a double threat in g and e and so wins by Lemma 2. If White plays in j instead of v, the argument is similar.

If Black plays in v, the proof that White has to play in s is similar.   □

The (1-2)-vertex and (0,2)-vertex gadgets. A $(1, 2)$-vertex gadget receives one ending point of a wire (Fig. 4a). If a stone is played on this ending point, it should offer the choice to defend either by playing in the starting point of a second wire, or by playing in the starting point of a third wire. That simulates a vertex with one edge going in and two edges going out. The $(0, 2)$-vertex gadget (or *starting-vertex* gadget) can be seen as a $(1, 2)$-vertex gadget where a stone has already been played on the ending point of the in-edge. The $(0, 2)$-vertex gadget represents the two possible choices of the first player at the beginning of the game.

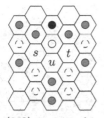

(a) The $(1,2)$-vertex gadget. The wire for the in-edge ends in u.

(b) The $(0,2)$-vertex gadget representing the starting vertex v_0.

Fig. 4. In these choice gadgets, White can defend by playing in s or in t. A North-West BW-wire starts in s and a North-East BW-wire starts in t.

Lemma 6. *If Black plays in the starting point u of a $(1,2)$-vertex gadget (Fig. 4a), and White does not play a threat, White is forced to play in one of the two ending points s and t. Then, if Black does not answer by a threat, they have to play in the other ending point.*

Proof. Black plays in u. Assume White plays neither in s, nor in t, nor a threatening move. Then Black plays in s. By Lemma 1, White has to play in t but Black wins by playing in the ending point of the wire starting at s by Lemma 3.

Assume White's answer to u is to play in s. t can now be seen as the ending point of the in-wire, so Black needs to play in t or make a threat by Lemma 3.□

Corollary 1. *If White is forced to play a threat or to open the game in one of the two opening points s and t of the $(0,2)$-vertex gadget (Fig. 4b). Then, if Black does not play a threat, they are forced to play in the other opening point.*

2.4 Assembling the Gadgets Together

Let (G, v_0) be a simplified instance of GG, and n be its number of vertices. G being bipartite, we denote by V_1 the side of the partition containing v_0, and V_2 the other side. Player 1 moves the token from vertices of V_1 to vertices of V_2 and player 2 moves the token from V_2 to V_1. We denote by ϕ the reduction from GG to HAVANNAH. Let us describe the construction of $\phi((G, v_0))$. As an example, we provide the reduction from the GG instance from Fig. 1 in Fig. 5.

The initial vertex v_0 is encoded by the gadget displayed in Fig. 4b. Each player 1's $(2,1)$-vertex is encoded by the $(2,1)$-vertex gadget of Fig. 3a, and each player 2's $(2,1)$-vertex is encoded by the same gadget in reverse color. Each player 1's $(1,2)$-vertex is encoded by the $(1,2)$-vertex gadget of Fig. 4a, and each player 2's $(1,2)$-vertex is encoded by the same gadget in reverse color.

All White's vertex gadgets are aligned and all Black's vertex gadgets are aligned on a parallel line. Whenever (u, v) is an edge in G, we connect an exit of the vertex gadget representing u to an entrance of a gadget encoding v using wires and crossover gadgets. Let n be the number of vertices in G, since G is of degree 3, we know that the number of edges is at most $3n/2$. The minimal size

Fig. 5. HAVANNAH gadgets representing the GG instance from Fig. 1.

in terms of HAVANNAH cells for a smallest wire and the size of a crossover are constants. Therefore the distance between Black's line and White's line is linear in n. Note that, two wires of opposite colors might be needed to connect two vertex gadgets or a vertex gadget and a crossover. Similarly, we can show that the distance between two vertices on Black's line or on White's line is constant.

Lemma 7. *If Black reenters a White's $(2,1)$-vertex gadget (Fig. 3b), and Black has no winning sequence of threats elsewhere, White wins.*

Proof. If Black reenters a White's $(2,1)$-vertex by playing in v, White plays in e. As Black cannot initiate a winning sequence, whatever Black then plays White can defend until Black is not threatening anymore. Then White plays in k or in l with a decisive double threat in m and n. □

Theorem 1. HAVANNAH *is* PSPACE-*complete*.

3   TwixT

Alex Randolph's TwixT is one of the most popular connection games. It was invented around 1960 and was marketed as soon as in 1962 [13]. In his book devoted to connection games, Cameron Browne describes TwixT as one of the most popular and widely marketed of all connection games [5]. We now briefly describe the rules of TwixT and refer to Moesker's master's thesis for an introduction and a mathematical approach to the strategy, and the description of a possible implementation [19].

TWIXT is a 2-player connection game played on a GO-like board. At their turn, player White and Black place a pawn of their color in an unoccupied place. Just as in HAVANNAH and HEX, pawns cannot be taken, moved, nor removed. When 2 pawns of the same color are spaced by a knight's move, they are linked by an edge of their color, unless this edge would cross another edge. At each turn, a player can remove some of their edges to allow for new links. The goal for player White (Black, respectively) is to link top and bottom (left and right, respectively) sides of the board. Note that sometimes, a player could have to choose between two possible edges that intersect each other. The *pencil and paper* version TWIXTPP where the edges of a same color are allowed to cross is also famous and played online.

As the length of a game of TWIXT is polynomially bounded, exploring the whole tree can be done with polynomial space using a minimax algorithm. Therefore TWIXT is in PSPACE.

Mazzoni and Watkins have shown that 3-SAT could be reduced to single-player TWIXT, thus showing NP-completeness of the variant [18]. While it might be possible to try and adapt their work and obtain a reduction from 3-QBF to standard two-player TWIXTT, we propose a simpler approach based on HEX. The PSPACE-completeness of HEX has already been used to show the PSPACE-completeness of AMAZONS, a well-known territory game [11].

We now present how we construct from an instance G of HEX an instance $\phi(G)$ of TWIXT. We can represent a cell of HEX by the 9×9 TWIXT gadgets displayed in Fig. 6. Let n be the size of a side of G, Fig. 7 shows how a TWIXT board can be paved by n^2 TWIXT cell gadgets to create a HEX board.

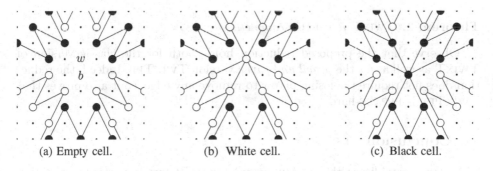

(a) Empty cell.   (b) White cell.   (c) Black cell.

Fig. 6. Basic gadgets needed to represent cells.

It is not hard to see from Fig. 6a that in each gadget of Fig. 7, move w (resp. b) is dominating for White (Black, respectively). That is, playing w is as good for White as any other move of the gadget. We can also see that the moves that are not part of any gadget in Fig. 7 are dominated for both players. As a result, if player Black (White, respectively) has a winning strategy in G, then player Black has a winning strategy in $\phi(G)$. Thus, G is won by Black if and only if $\phi(G)$ is won by Black. Therefore determining the winner in TWIXT is at least as hard as in HEX, leading to the desired result.

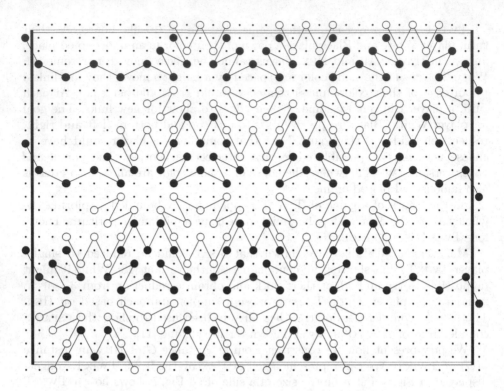

Fig. 7. Empty 3×3 HEX board reduced to a TWIXT board.

Theorem 2. TWIXT *is* PSPACE-*complete.*

Observe that the proposed reduction holds both for the classic version of TWIXT as well as for the *pencil and paper* version TWIXTPP. Indeed, the reduction does not require the losing player to remove any edge, so it also proves that TWIXTPP is PSPACE-hard.

4   Conclusion

This paper establishes the PSPACE-completeness of two important connection games, HAVANNAH and TWIXT. The proof for TWIXT is a reduction from HEX and applied to TWIXTPP. The proof for HAVANNAH is more involved and is based on the GENERALIZED GEOGRAPHY problem restricted to bipartite graphs of degree 3. This HAVANNAH reduction only used the loop winning condition, but it is easy to show that HAVANNAH without the loop winning condition can simulate HEX and is PSPACE-hard as well. For both reductions, the size of the resulting game is only linearly larger than the size of the input instance.

The complexity of other notable connection games remains open. In particular, the following games seem to be good candidates for future work on the complexity of connection games.

In LINES OF ACTION, each player starts with two groups of pieces and tries to connect all their pieces by moving these pieces and possibly capturing opponent pieces [24]. While the goal of LINES OF ACTION clearly makes it a connection game, the mechanics distinguishes it from more classical connection games as no pieces are added to the board and existing pieces can be moved or removed. As a result, it is not even clear that LINES OF ACTION is in PSPACE.

SLITHER is closer to HEX but each move actually consists of putting a new stone on the board and possibly moving another one. Obtaining a PSPACE-hardness result for SLITHER is not so easy since the rules allow a player to influence two different areas of the board in a single turn.

References

1. Anshelevich, V.V.: A hierarchical approach to computer Hex. Artif. Intell. **134** (1–2), 101–120 (2002)
2. Arneson, B., Hayward, R.B., Henderson, P.: Monte Carlo tree search in Hex. IEEE Trans. Comput. Intell. AI Games **2**(4), 251–258 (2010)
3. Björnsson, Y., Hayward, R., Johanson, M., van Rijswijck, J.: Dead cell analysis in Hex and the Shannon game. In: Bondy, A., Fonlupt, J., Fouquet, J.-L., Fournier, J.-C., Alfonsín, J.L.R. (eds.) Graph Theory in Paris, pp. 45–59. Springer-Birkhäuser, Basel (2007)
4. Bonnet, É., Jamain, F., Saffidine, A.: On the complexity of trick-taking card games. In: Rossi, F. (ed.) 23rd International Joint Conference on Artificial Intelligence (IJCAI), Beijing, China, August 2013. AAAI Press
5. Browne, C.: Connection Games: Variations on a Theme. AK Peters, Natick (2005)
6. Browne, C., Powley, E., Whitehouse, D., Lucas, S., Cowling, P., Rohlfshagen, P., Tavener, S., Perez, D., Samothrakis, S., Colton, S.: A survey of Monte Carlo tree search methods. IEEE Trans. Comput. Intell. AI Games **4**(1), 1–43 (2012)
7. Bruno, J., Weinberg, L.: A constructive graph-theoretic solution of the Shannon switching game. IEEE Trans. Circ. Theory **17**(1), 74–81 (1970)
8. Demaine, E.D., Hearn, R.A.: Playing games with algorithms: algorithmic combinatorial game theory. In: Nowakowski, R.J. (ed.) Games of No Chance III, pp. 3–56. Cambridge University Press, Cambridge (2009)
9. Even, S., Tarjan, R.E.: A combinatorial problem which is complete in polynomial space. J. ACM (JACM) **23**(4), 710–719 (1976)
10. Ewalds, T.: Playing and solving Havannah. Master's thesis, University of Alberta (2012)
11. Furtak, T., Kiyomi, M., Uno, T., Buro, M.: Generalized Amazons is PSPACE-complete. In: Kaelbling, L.P., Saffiotti, A. (eds.) 19th International Joint Conference on Artificial Intelligence (IJCAI), pp. 132–137 (2005)
12. Hearn, R.A., Demaine, E.D.: Games, Puzzles, and Computation. AK Peters, Natick (2009)
13. Huber, J.R.: Computer game playing: the game of Twixt. Master's thesis, Rochester Institute of Technology (1983)
14. Kiefer, S.: Die Menge der virtuellen Verbindungen im Spiel Hex ist PSPACE-vollständig. Studienarbeit Nr. 1887, Universität Stuttgart, July 2003
15. Lichtenstein, D., Sipser, M.: Go is polynomial-space hard. J. ACM (JACM) **27**(2), 393–401 (1980)

16. Lorentz, R.J.: Improving Monte–Carlo tree search in Havannah. In: van den Herik, H.J., Iida, H., Plaat, A. (eds.) CG 2010. LNCS, vol. 6515, pp. 105–115. Springer, Heidelberg (2011)
17. Maarup, T.: Hex: everything you always wanted to know about Hex but were afraid to ask. Master's thesis, Department of Mathematics and Computer Science, University of Southern Denmark, Odense, Denmark (2005)
18. Mazzoni, D., Watkins, K.: Uncrossed knight paths is NP-complete (1997). http://www.cs.cmu.edu/~kw/pubs/Twixt_Proof_Draft.txt
19. Moesker, K.: Twixt: theory, analysis and implementation. Master's thesis, Maastricht University (2009)
20. Reisch, S.: Hex ist PSPACE-vollständig. Acta Inform. **15**, 167–191 (1981)
21. Schaefer, T.J.: On the complexity of some two-person perfect-information games. J. Comput. Syst. Sci. **16**(2), 185–225 (1978)
22. Teytaud, F., Teytaud, O.: Creating an upper-confidence-tree program for Havannah. In: van den Herik, H.J., Spronck, P. (eds.) ACG 2009. LNCS, vol. 6048, pp. 65–74. Springer, Heidelberg (2010)
23. Uiterwijk, J., Moesker, K.: Mathematical modelling in Twixt. In: Löwe, B. (ed.) Logic and the Simulation of Interaction and Reasoning, pp. 29 (2009)
24. M.H.M, Winands, Björnsson, Y., Saito, J.-T.: Monte Carlo tree search in lines of action. IEEE Trans. Comput. Intell. AI Games **2**(4), 239–250 (2010)

Material Symmetry to Partition Endgame Tables

Abdallah Saffidine[1]([⊠]), Nicolas Jouandeau[2], Cédric Buron[1,3],
and Tristan Cazenave[1]

[1] LAMSADE, Université Paris-Dauphine, Paris, France
abdallahs@csc.unsw.edu.au
[2] LIASD, Université Paris 8, Saint-Denis, France
[3] National Chiao Tung University, Hsinchu, Taiwan

Abstract. Many games display some kind of *material symmetry*. That is, some sets of game elements can be exchanged for another set of game elements, so that the resulting position will be equivalent to the original one, no matter how the elements were arranged on the board. Material symmetry is routinely used in card game engines when they normalize their internal representation of the cards.

Other games such as CHINESE DARK CHESS also feature some form of material symmetry, but it is much less clear what the normal form of a position should be. We propose a principled approach to detect material symmetry. Our approach is generic and is based on solving multiple relatively small sub-graph isomorphism problems. We show how it can be applied to CHINESE DARK CHESS, DOMINOES, and SKAT.

In the latter case, the mappings we obtain are equivalent to the ones resulting from the standard normalization process. In the two former cases, we show that the material symmetry allows for impressive savings in memory requirements when building endgame tables. We also show that those savings are relatively independent of the representation of the tables.

1 Introduction

Retrograde analysis is a tool to build omniscient endgame databases. It has been used in CHESS to build endgame databases of up to six pieces [14,19,20]. It has also been used in a similar way to build endgame databases for CHECKERS [17] that have helped solving the game [18]. It has also been successfully used for strongly solving the game of AWARI with a parallel implementation that evaluated all the possible positions of the game [15]. Other games it has helped solving are FANORONA [16] and NINE MEN'S MORRIS [8]. Retrograde analysis has also been applied to CHINESE CHESS [6], KRIEGSPIEL [4], and GO [2].

An important limitation on the use of endgame tables is the size needed to store all the computed results. Consequently, elaborate compression techniques have been proposed and they have been instrumental in achieving some of the aforementioned results [17]. An orthogonal approach to alleviate the memory bottleneck is to use symmetries to avoid storing results that can be deduced

H.J. van den Herik et al. (Eds.): CG 2013, LNCS 8427, pp. 187–198, 2014.
DOI: 10.1007/978-3-319-09165-5_16, © Springer International Publishing Switzerland 2014

from already stored results by a symmetry argument. There are two kinds of such symmetry arguments. *Geometrical symmetry* is the simplest case [5]: for instance if a CHESS position p does not feature any pawn and castling rights have been lost then any combination of the following operations yields a position equivalent to p. Flipping pieces along a vertical axis between the 4th and the 5th columns, flipping pieces along a horizontal axis between the 4th and the 5th rows, or rotating the board. *Material symmetry* is typical from card games, in particular when suits play an equivalent role. In most trick-taking games, for instance, if a position p only contains Hearts, any position obtained by replacing all cards with cards of same rank from an other suit is equivalent.

State-of-the-art engines for card games such as SKAT or BRIDGE already use material symmetry detection in their transposition and endgame tables. However, recognising the most general form of material symmetry is not as straightforward in some other games, notably CHINESE DARK CHESS. In this paper, we propose a principled framework to detect material symmetry and show how it can be applied to three games: CHINESE DARK CHESS, DOMINOES, and SKAT. At the core of our method lies the sub-graph isomorphism problem which has been extensively studied in computer science [12, 21].

CHINESE DARK CHESS is a popular game in Asia [3]. One of the key features of endgame databases in CHINESE DARK CHESS is that some combinations of pieces are equivalent. It means that in some endgames, a piece can be replaced by another piece without changing the result of the endgame. Therefore an endgame computed with the first piece can be used as is for the endgame positions containing the other piece in place of the first piece. This property reduces the number of endgame databases that have to be computed. Using the relative ordering of pieces instead of the exact values is similar to partition search in BRIDGE that store the relative ordering of cards in the transposition table instead of the exact values [9].

DOMINOES is also a popular game and in the endgame some of the values of pieces can be replaced by other values without changing the outcome of the game. We use this property to compute a reduced number of endgame tables for the perfect information version of the game.

SKAT is a popular game in Germany. Here again in the endgame, some cards can be replaced by other cards. We present in this paper the memory reduction that may be expected from using this property to compute complete information endgame tables. Perfect information endgame tables are important in SKAT since the Monte-Carlo approach in SKAT consists in solving many perfect information versions of the current hand [1, 11, 13]. Using endgame tables enables to speed-up the solver.

The second section describes CHINESE DARK CHESS, DOMINOES, and SKAT. We then show in Sect. 3 how material interaction graphs can be constructed for these domains. Section 4 recalls the principles of endgame table construction. The fifth section gives experimental results, and the last section concludes.

2   Domains Addressed

Below we briefly describe the three domains addressed, viz. CHINESE DARK
CHESS (Sect. 2.1), DOMINOES (Sect. 2.2), and SKAT (Sect. 2.3).

2.1   Chinese Dark Chess

CHINESE DARK CHESS is a stochastic perfect information game that involves
two players on 4×8 rectangular board. Each player starts with one king, two
guards, two bishops, two knights, two rooks, two cannons, and five pawns that
are the same pieces as in CHINESE CHESS. The pieces can be denoted by Chinese
characters or numbers as shown in Table 1. Pieces can move vertically and horizontally from one square to an adjacent free square. Captures are performed on
vertical and horizontal adjacent squares except for cannons that capture pieces
by jumping over another piece. Such a jump is done over a piece (called the
jumping piece) and on a piece (called the *target piece*). Free spaces can stand
between its initial position and the jumping piece and between the jumping piece
and the target position. A piece can capture another piece according to the possibilities mentioned in Table 1. Considering number representation, pieces can
capture lower or equal numbers (except the king that cannot capture any pawn,
and except the cannon that can capture any piece).

Table 1 summarizes alternative pieces' names, each side's icons and possible
captures. The column titled "Capture rule" lists the pieces that a piece can
capture.

Table 1. Pieces representation and capture rules in CHINESE DARK CHESS.

Names	Representation	Capture rule	Note
King (K)	帥 將 , ⑦ ❼	all opponent piece except pawns	
Guard (G)	仕 士 , ⑥ ❻	all inferior opponent pieces	
Bishop (B)	相 象 , ⑤ ❺	all inferior opponent pieces	
Knight (N)	傌 馬 , ④ ❹	all inferior opponent pieces	
Rook (R)	俥 車 , ③ ❸	all inferior opponent pieces	
Cannon (C)	炮 炮 , ② ❷	all opponent pieces	jump capture
Pawn (P)	兵 卒 , ① ❶	opponent king and pawns	

At the beginning, pieces are randomly placed on the board, facing down.
Typical positions thus allow 3 types of moves, flipping an unknown piece, moving
a piece to an adjacent free square, capturing an opponent piece.

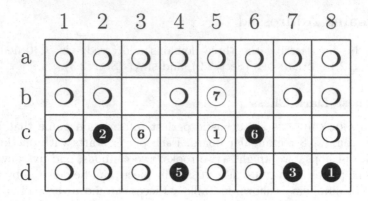

Fig. 1. Sample CHINESE DARK CHESS position after 10 turns of play.

If a player runs out of legal moves, possibly because all their pieces have been captured, the game ends and that player loses the game. The game ends with a draw if no capture is made for 40 plies.

Figure 1 shows the resulting board situation after the following 10 turns (Parentheses indicate revealed piece for flipping moves. Moving and capturing moves indicate two coordinates. Unknown pieces are represented by white circles.): b5(k) d8(P); d7(R) d4(p); c7(G) c6(c); c4(C) c4-c6; c5(p) c3(B); b6(r) c3-c4; b6-c6 b7-b6; c2(C) c4-d4; b3(g) d4-c4; b3-c3 c4-d4; It is now the first player's turn to play. The first player is white. First player possible moves are: b5-b6; c3-b3; c3-c2; c3-d3; c5-c4. Second player possible moves are: c2-c5; c6-c5; c6-b6; c6-c7; d4-c4; d7-c7.

Sometimes, the outcome of the game depends only on the player's turn, even if a player has a stronger piece, as shown in Fig. 2. If it is black to play, then white wins (for example: b2-b1 c3-c2; b1-a1 c2-b2; a1-a2 b2-a2; white wins).

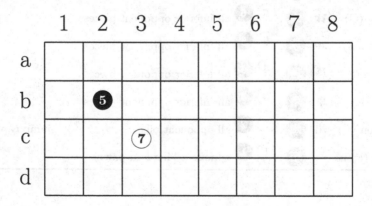

Fig. 2. Sample CHINESE DARK CHESS endgame position where the outcome depends on who's next turn to play.

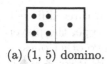
(a) (1, 5) domino.

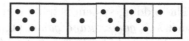
(b) Chain of three tiles after a player has played (3, 2).

Fig. 3. A tile and a chain of matching tiles in DOMINOES.

If it is white to play, then it is a draw game (for example: c3-c2 b2-b3; c2-b2 b3-c3; b2-b3 c3-c2; b3-c3 c2-b2... until draw).

2.2 The Game of Dominoes

The game of DOMINOES is an ancient game, played with rectangular tiles called dominoes. We do not know precisely when or where the game appeared, but it is widely played all around the world.

The tiles of the game are rectangular and are divided into two sides. Each side represents a number of points between 0 (empty) and 6, as shown in Fig. 3a. They are called *dominoes* or simply tiles.

In this paper, the tiles will be represented by the two numbers that are written on it. For instance the domino in Fig. 3a is represented by $(1, 5)$. The tile (i, j) is the same as the tile (j, i) and only appears once in the set. Thus, there are 28 tiles: $(0, 0), (0, 1), \ldots, (0, 6), (1, 1), (1, 2), \ldots, (6, 6)$.

The board of the game of DOMINOES consists of a chain of dominoes already laid down. At the beginning of the game, the first player puts his "strongest" domino (the one which has the most points) on the table and starts the chain. Then, the players take turn laying down one tile they possess according to the following matching constraint. To play a tile, it must have at least one side with the same number of points of one of the tiles at the end of the chain on the table (Fig. 3). If a player cannot play a tile, they draw a tile from the stock or pass when there are no stock tiles remaining.

If a player has laid down all their tiles, or if the stock is empty and no player can play, the game ends and the player with fewer points in hand wins.

While there are multi-player variants of the game of DOMINOES, we focus in our experiments on the two-player case. Additionally, since we focus on the endgames, we assume that the stock is empty and, as a consequence, that the game is a game of perfect information. Our method to detect material symmetry extends easily to other settings.

2.3 SKAT

SKAT is a card game for three players. SKAT is played with 32 cards using 8 ranks (7–10, Jack, Queen, King, and Ace) and 4 suits (Club ♣, Spade ♠, Heart ♡, and Diamond ◇). Each card is associated to a suit and a rank, for instance, we have the Jack of Heart (♡J) or the Ten of Spade (♠T). A game of SKAT is divided into a bidding and a playing phase. Before bidding, each player receives 10 cards and the two remaining ones are placed on the center, and called the *Skat*.

Bidding. The bidding system is quite complex and can lead to three different types of game: *suit game, grand game,* or *null game.*[1] In each type, the player who made the highest bid gets the skat and plays against the two other ones.

In the *suit game,* a suit is chosen as trump. The order of cards is so: ♣J, ♠J, ♡J, ♢J, then the trump suit: A, T, K, Q, 9, 8, 7, then each of the remaining suits, in the same order.

In the *grand game,* there is no trump suit, but the Jacks are still considered as trumps. The order for the other cards is the same as in the suit game.

In the *null game,* there is no trump. The order of the cards is not the same as in the other games: A, K, Q, J, T, 9, 8, 7.

Playing. The playing part of the game is a succession of tricks. Each trick, the player who won the previous trick plays first, and decides which suit is played in the trick. The other players have to play this suit. If a player cannot play the suit the first player played, he must play a trump. If he[2] has neither the first suit played nor trump, he can play any other suit.

If there is a trump in the trick, the highest trump wins the trick. If there is not, the highest card of the suit that the first player played wins. The player who won the trick put the cards of the trick behind him and begins the next one.

Counting the Points. When all the cards have been played, each team count their points. The team which got most of the points wins the round. The points are distributed as follows. A: 11 points, T: 10 points, K: 4 points, Q: 3 points, J: 2 points, 9, 8, and 7: 0 points

3 Detecting Material Symmetry

Multiple games feature classes of endgames that are globally equivalent one to another. For instance, in a trick-taking card game such as BRIDGE, the class of 8 cards endgames where all cards are Hearts is equivalent to the class of 8 cards endgames where all cards are Spades, and both are equivalent to that where all cards are Clubs and to that with Diamonds. Indeed, we can exhibit a mapping from one class C to an equivalent class C_0 that will associate to each position $p \in C$ a position $p_0 \in C_0$ such that the score and optimal strategy in p can easily be deduced from the score and optimal strategy in p_0.

As a result, it is only necessary to build one endgame database for each equivalence class of endgames, rather than one for each class of endgames. A first approach to avoid building unnecessary endgame databases is to proceed to some form of normalization. However, the normalization process to be adopted is not always trivial. We propose here a principled method to find representatives of the equivalence classes.

---

[1] For more details, we refer to the rules from the International Skat Players Association: http://www.ispaworld.org/downloads/ISkO-rules-2007-Canada.pdf.

[2] For brevity, we use 'he' and 'his' whenever 'he or she' and 'his or her' are meant.

The basic idea is to represent interactions between pieces in a graph. The interaction can be specific to a game. For example, in CHINESE DARK CHESS the interaction is a capture whereas it is a connection in DOMINOES. Figure 4 gives the graph for the pieces of CHINESE DARK CHESS. There is an arrow from one piece to another if the first piece can capture the second one. Figure 6 gives the graph for DOMINOES and Fig. 5 gives the graph for SKAT assuming a suit game with Clubs as trump.

Once the graph has been built for a game, it can be used to detect equivalent endgames. The principle is to extract the sub-graph of the pieces of a first endgame and to compare it with the sub-graph of the second endgame. The comparison consists in finding if the two sub-graphs are isomorphic. Subgraph isomorphism is a hard problem, but for the games that we address a naive algorithm is fast enough to compare sub-graphs of interest [21].

4 Endgame Tables

Retrograde analysis is completed in two steps. The first step consists in enumerating all winning positions for one player. The second step consists in repeatedly

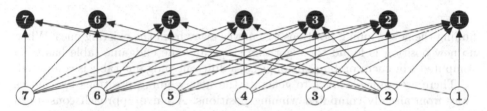

Fig. 4. Capture relationship in CHINESE DARK CHESS. The allowed captures for black were not represented so as to avoid cluttering the graph.

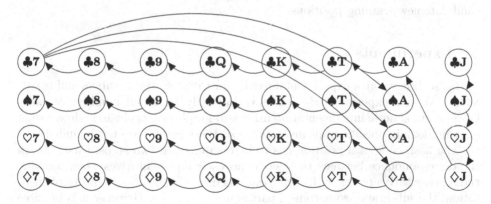

Fig. 5. Trick winning relationship for cards in SKAT assuming Clubs (♣) are trumps. To avoid cluttering, not all edges are drawn. The full graph is obtained by taking the transitive closure of the graph represented.

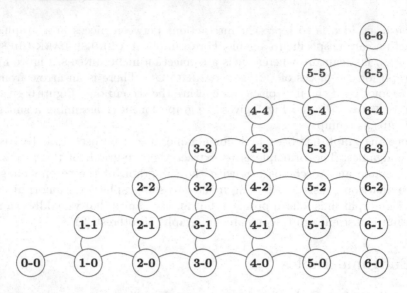

Fig. 6. Playability relationship for tiles in DOMINOES. To avoid cluttering, not all edges are drawn. The full graph is obtained by adding edges between every tile in a given column as well as between every tile in a given row.

finding the new winning positions two moves away from the existing ones. When no new position is found the algorithm stops and the endgame table has been computed. For each combination of pieces a different endgame table is computed.

There are multiple ways to generate the new winning positions two moves away from already computed winning positions. A naive approach consists in running through all possible positions and checking if they are two moves away from already computed ones. A more elaborate approach consists in doing unmoves from already computed winning positions to find a restricted set of candidate new winning positions.

5   Experiments

In this section we give experimental results for CHINESE DARK CHESS and DOMI-NOES. We also experimented with SKAT but only for small numbers of cards because our generic implementation using sub-graph isomorphism is slower than existing domain specific implementations, so these results are not included.

The hash code of a position has to be computed in two steps. First, obtain the correspondence between the pieces and their representatives, then combine the representative pieces with their locations into a hash code. In our implementation, the sub-graph isomorphism part is not optimized. However it is possible to pre-compute all sub-graph isomorphism between tables and to have a fast correspondence between a table and its representative.

5.1   Equivalence Classes

Assume that the endgame table is split into multiple tables depending on the number of game elements (pieces/tiles/cards). Assume also that the resulting tables are split further according to the sets of elements that constitute the positions. Such a decomposition is natural and makes distributing the workload easier as there is no dependency between different sets of elements if they have the same number of elements [10]. Thus, each set of game elements has a corresponding endgame table which can be stored in its own file.

We begin by comparing the number of possible files without using the reduction and the number of remaining files with our method. Material symmetry allows us to reduce significantly the number of needed files. For instance, for CHINESE DARK CHESS with 4 elements, there are 1737 sets of elements that are consistent with the rules of the game. However many such sets are equivalent, and if we detect material symmetry, we can reduce the number of needed sets down to 186, which represent a reduction factor of 9.34. Table 2 shows the number of sets, the number of sets needed when material symmetry is used, and the resulting reduction factor for a given number of game elements in CHINESE DARK CHESS and DOMINOES.

5.2   Perfect Hashing Representation

The reduction factors presented in Table 2 do not necessarily match with the memory actually needed to store the endgame tables. Indeed, different sets of pieces might lead to tables of different size and equivalence classes might have different number of elements. For instance, in CHINESE DARK CHESS the Kkm file corresponds to 29,760 positions, while the kPP file corresponds only to 14,880 positions because the two pawns are indistinguishable. In this section, we assume that we store the result for every position in the corresponding file using a perfect hashing function, or index.

Table 2. Total number of tables and number of representatives.

Game	Elements	Total	Representatives	Reduction factor
CHINESE DARK CHESS	2	49	8	6.13
	3	378	46	8.22
	4	1737	186	9.34
	5	5946	672	8.92
	6	16,524	2240	7.38
	7	39,022	6694	5.83
	8	80,551	17,662	4.56
DOMINOES	3	4032	20	201.6
	4	30,303	61	496.8
	5	180,180	185	973.9
	6	868,140	563	1542.0

Table 3. Size of the endgame database with a perfect hashing representation.

Game	Elements	Total Positions	Memory	Representatives Positions	Memory	Reduction factor
CHINESE DARK	2	4.961×10^4	2.97 KB	7.936×10^3	496 B	6.13
CHESS	3	9.999×10^6	610.3 KB	1.131×10^6	69.02 KB	8.84
	4	1.140×10^9	67.92 MB	1.142×10^8	6.81 MB	9.92
	5	9.036×10^{10}	5.26 GB	1.013×10^{10}	603.8 MB	8.92
	6	5.440×10^{12}	316.7 GB	8.002×10^{11}	46.58 GB	6.80
	7	2.601×10^{14}	14.78 TB	5.247×10^{13}	2.98 TB	4.96
	8	1.014×10^{16}	576.1 TB	2.756×10^{15}	156.7 TB	3.68
DOMINOES	3	21,168	2.58 KB	92	11.50 B	230.09
	4	550,368	67.18 KB	996	124.50 B	552.58
	5	8,026,200	979.76 KB	7,854	981.75 B	1021.93
	6	82,556,550	9.84 MB	53,790	6.57 KB	1541.99

We did not use any advanced compression mechanism [17]. For instance, CHINESE DARK CHESS has 2 geometrical symmetries and we used 2 bits to encode whether a position was won, lost, or draw. Therefore, for each file we need half as many bits as the number of positions.

Table 3 shows the size needed for endgame tables with various number of elements using a perfect hashing representation. In this table, we added the different possible positions of each file for a given number of elements with and without the reduction, and then calculated the real reduction factor. Notice that the reduction factor in Table 3 are close to those in Table 2 but are not always the same.

5.3 Won Positions

The following alternative representation for endgame tables can be used. We will show that reduction factors are similar under this representation as well. For a given set of pieces, store the list of won positions. We can then also detect that a position is lost if the position with reversed colors is in the table. Finally, a position is a draw if it and the reverse position are not in table. Again, we can split the endgame table according to the set of pieces that constitute it and use one file for each possible set.

We have computed endgames tables with only won positions. Table 4 gives the total number of won positions that would need to be stored for a given number of game elements. It also displays the number of positions that need to be stored if we use material symmetry and only store one representative file for each equivalence class. The last column indicate the savings in terms of number of stored positions when material symmetry is taken into account.

6   Conclusion

We have presented a general method based on sub-graph isomorphism that allows detection material symmetries. We have shown that it could be applied to SKAT,

Table 4. Size of the endgame database when only storing won positions.

Game	Elements	Total	Representatives	Reduction factor
CHINESE DARK CHESS	2	6448	744	8.67
	3	1,650,763	171,516	9.62
	4	156,204,805	15,418,377	10.13
DOMINOES	3	22,785	104	219.09
	4	409,234	747	547.84
	5	5,706,631	5,741	994.01

DOMINOES, and CHINESE DARK CHESS. Detecting material symmetry makes it easier to build larger endgame databases. While finding a representative for positions that are equivalent under material symmetry (that is, defining a normal form) for SKAT and DOMINOES is relatively easy, the CHINESE DARK CHESS case is more intricate as the relationship between the different pieces is more elaborate. Our approach solves this problem for CHINESE DARK CHESS, too, and allows a reduction factor between 5 and 10 in the size needed for storing the endgame tables.

Note that although we do not report any time measurement, a significant reduction factor can also be expected for the time needed to compute the endgame tables. In our framework, we propose to store a file mapping sets of pieces to their representative. This file is much smaller than the corresponding endgame tables. Thus, the only runtime cost induced by our method is an additional indirection when looking up a position in the database.

A generalisation of transposition tables has been proposed by Furtak and Buro in order to regroup cases that have similar structures but different pay-offs [7]. Determining to which extent the proposed method for detecting material symmetry can be extended to account for payoff similarity seems to be a promising line for future work.

References

1. Buro, M., Long, J.R., Furtak, T., Sturtevant, N.: Improving state evaluation, inference, and search in trick-based card games. In: 21st International Joint Conference on Artificial Intelligence (IJCAI 2009) (2009)
2. Cazenave, T.: Generation of patterns with external conditions for the game of go. Adv. Comput. Games **9**, 275–293 (2001)
3. Chen, B.-N., Shen, B.-J., Hsu, T.: Chinese dark chess. ICGA J. **33**(2), 93 (2010)
4. Ciancarini, P., Favini, G.P.: Solving kriegspiel endings with brute force: the case of KR vs. K. In: van den Herik, H.J., Spronck, P. (eds.) ACG 2009. LNCS, vol. 6048, pp. 136–145. Springer, Heidelberg (2010)
5. Culberson, J.C., Schaeffer, J.: Pattern databases. Comput. Intell. **14**(3), 318–334 (1998)
6. Fang, H., Hsu, T., Hsu, S.-C.: Construction of Chinese Chess endgame databases by retrograde analysis. In: Marsland, T., Frank, I. (eds.) CG 2001. LNCS, vol. 2063, pp. 96–114. Springer, Heidelberg (2002)

7. Furtak, T., Buro, M.: Using payoff-similarity to speed up search. In: 22nd International Joint Conference on Artificial Intelligence (IJCAI2011), pp. 534–539. AAAI Press (2011)
8. Gasser, R.: Solving nine men's morris. Comput. Intell. **12**(1), 24–41 (1996)
9. Ginsberg, M.L.: Partition search. In: National Conference On Artificial Intelligence (AAAI1996), pp. 228–233 (1996)
10. Goldenberg, M., Lu, P., Schaeffer, J.: TrellisDAG: a system for structured DAG scheduling. In: Feitelson, D.G., Rudolph, L., Schwiegelshohn, U. (eds.) JSSPP 2003. LNCS, vol. 2862, pp. 21–43. Springer, Heidelberg (2003)
11. Kupferschmid, S., Helmert, M.: A skat player based on Monte-Carlo simulation. In: van den Herik, H.J., Ciancarini, P., Donkers, H.H.L.M.J. (eds.) CG 2006. LNCS, vol. 4630, pp. 135–147. Springer, Heidelberg (2007)
12. Kuramochi, M., Karypis, G.: Frequent subgraph discovery. In: 2001 Proceedings of IEEE International Conference on Data Mining, ICDM 2001, pp. 313–320. IEEE (2001)
13. Long, J.R.: Search, inference and opponent modelling in an expert-caliber skat player, PhD thesis, University of Alberta (2011)
14. Nalimov, E.V., Haworth, G.M., Heinz, E.A.: Space-efficient indexing of chess endgame tables. ICGA J. **23**(3), 148–162 (2000)
15. Romein, J., Bal, H.E.: Solving awari with parallel retrograde analysis. Computer **36**(10), 26–33 (2003)
16. Schadd, M.P.D., Winands, M.H.M., Uiterwijk, J.W.H.M., Van Den Herik, H.J., Bergsma, M.H.J.: Best play in fanorona leads to draw. New Math. Nat. Comput. **4**(3), 369–387 (2008)
17. Schaeffer, J., Björnsson, Y., Burch, N., Lake, R., Sutphen, S.: Building the checkers 10-piece endgame databases. In: Van Den Herik, H.J., Iida, H., Heinz, E.A. (eds.) Many Games, Many Challenges. IFIP, vol. 135, pp. 193–210. Springer, Heidelberg (2004)
18. Schaeffer, J., Burch, N., Björnsson, Y., Kishimoto, A., Müller, M., Lake, R., Paul, L., Sutphen, S.: Checkers is solved. Science **317**(5844), 1518–1522 (2007)
19. Thompson, K.: Retrograde analysis of certain endgames. ICCA J. **9**(3), 131–139 (1986)
20. Thompson, K.: 6-piece endgames. ICCA J. **19**(4), 215–226 (1996)
21. Ullmann, J.D.: An algorithm for subgraph isomorphism. J. ACM (JACM) **23**(1), 31–42 (1976)

Further Investigations of 3-Member Simple Majority Voting for Chess

Kristian Toby Spoerer$^{(\boxtimes)}$, Toshihisa Okaneya, Kokolo Ikeda, and Hiroyuki Iida

Japan Advanced Institute of Science and Technology,
1-1 Asahidai, Nomi, Ishikawa 923-1292, Japan
{kristian,s1010014,kokolo,iida}@jaist.ac.jp

Abstract. The 3-member simple majority voting is investigated for the game of Chess. The programs STOCKFISH, TOGAII, and BOBCAT are used. Games are played against the strongest member of the group and against the group using simple majority voting. We show that the group is stronger than the strongest program. Subsequently, we investigate the research question, "under what conditions is 3-member simple majority voting stronger than the strongest member?" To answer this question we perform experiments on 27 groups. Statistics are gathered on the situations where the group outvoted the group leader. We found two conditions as an answer to the research question. First, group members should be almost equal in strength whilst still showing a small, but significant strength difference. Second, the denial percentage of the leaders candidate move depends on the strength of the members.

1 Introduction

Group superiority has been studied as early as 1898. Then, Triplett [7] focused on the motivating effect of groups. He found that a group of individuals each completing the same task would naturally start to compete. By the additional effort, Triplett voiced that the group would collectively perform better.

A similar kind of study was carried out in 1932, when Shaw [2] investigated groups of cooperating puzzle solvers. In these groups there was explicit communication of ideas. It was found that the groups seemed *"assured of a much larger proportion of correct solutions than individuals"*. It was proposed that the main cause was that mistakes made by individuals were being corrected by the group.

More recently, from 1985 onwards, a man-machine group for playing Chess was investigated by Althöfer [4]. In this kind of group, a human picked a move to play from a number of computer-proposed candidate moves. The resulting improvement came from combining the tactics and memory ability of the machines with the strategy and learning of the humans.

A pure machine group began to be investigated in 1991 [3], and afterwards by [5] and then [9]. These works analysed the simple majority voting method, a type of consultation algorithm. In this method, a group of computer programs each propose an action, and the majority proposal is accepted as the group decision.

H.J. van den Herik et al. (Eds.): CG 2013, LNCS 8427, pp. 199–207, 2014.
DOI: 10.1007/978-3-319-09165-5_17, © Springer International Publishing Switzerland 2014

In [3] a team was made from three Chess programs, experiments were done by hand, and results showed that the group could not beat the leader program, but could beat the other two programs. In [9] the same experiment was performed on Shogi and results showed that the team could beat all its members.

In [1], Cook investigated an alternative method for computing a group move for Go. The method, rather than using voting, used a group of human and computer players to analyse each others moves and demonstrate mistakes. This method showed an improvement over the strongest member of the group.

The majority algorithm is also used in Machine Learning. The weighted majority algorithm [6] can be used to combine independent opinions from several classifiers by taking the answer with the largest total weight. An incorrect final classification penalises the classifiers who agreed with that choice by reducing their weights. The algorithm learns which classifiers are more likely to be reliable and have a larger weight.

When the simple majority algorithm is applied to games, members are not motivated to do better by seeing an outperforming group member, candidate solutions are not proposed to be considered by the group, and the strengths of different kinds of group members are not combined. A group using majority gains extra strength from the combination of independently formed solutions. There is no 'parallelisation', which could contribute to an improvement as if a Chess program was decomposed and parallelised. In the majority algorithm separate programs produce a complete answer each, and these are combined.

In 2012 Sato et al. [10] performed a formal analysis of this improvement and concluded "*consultation algorithm with random numbers works well if and only if the expected improvement of consultation is greater than the expected reduction of consultation and the expected improvement of each engine is less than the expected reduction of each engine.*"

In this work we continue the experiments by (1) Althöfer [3] on Chess and (2) Obata et al. [9] on Shogi. We perform the same experiment as Althöfer, but we use different Chess programs, and we play 1,000 games per match as opposed to 20 games by Althöfer. We first test if majority voting using three separate Chess programs can play a better game than the leader program. We then investigate "*under what conditions does 3-member simple majority voting play a better game than the leader program?*"

The structure of the paper is as follows. Section 2 provides a description of the method that was used to construct a group and test its performance. Section 3 presents results from experiments on Chess, and Sect. 4 concludes the paper.

2   Method

Experiments were designed to analyse performance of groups of three Chess programs. STOCKFISH, TOGAII, and BOBCAT were used to make the group, see Table 1. The GUI used for playing Chess games was xBoard.

Table 1. Information about the Chess programs.

Program	Executable URL
STOCKFISH	stockfish-222-sse42-ja-intel
	http://stockfishchess.org/download/
TOGAII	toga141se-8cpu
	http://www.computerchess.info/downloads/engines/TogaII/TogaIIv1.4.1SE.7z
BOBCAT	bobcat_v3.25_x64
	https://github.com/Bobcat/bobcat/downloads
xBoard	winboard.exe (4.6.2)
	http://www.gnu.org/software/xboard/

2.1 Simple Majority Voting

The final group decision was computed using simple majority voting as described in [9]. The procedure is

1. Compute n candidate moves by searching with each of the n programs in the group.
2. Sum the total count of each candidate move.
3. If there is a majority candidate, select it as the group move.
4. If there is not a majority candidate, use the leader's (strongest member) proposed move to break the tie.

For 3 member Simple Majority Voting we set $n = 3$.

2.2 Group Composition

$D_P \in \mathbb{N}$ represents the depth-limit for program P when calculating a move for a given Chess position. This number is used by our implementation by passing "**go depth** D_P" to each program P in the group, as specified in the UCI protocol[1]. We use the notation (D_S, D_T, D_B) to represent a group of Chess programs. A group can then be described by, for example, $(D_S = 9, D_T = 8, D_B = 8)$ or simply $(9, 8, 8)$. This means that STOCKFISH searches to depth 9, TOGAII searches to depth 8, and BOBCAT searches to depth 8.

We formed four sets of groups, $S_1 = (8, 7, x)$, $S_2 = (9, 8, x)$, $S_3 = (10, 9, x)$ and $S_4 = (11, 10, x)$, such that $D_S - 5 \le x \le D_S + 1$, to make 28 groups in total. We assume that a program performs better as the search depth is increased, and also that each set is increasingly better.

We attempted to make groups with the following condition

$$\text{STOCKFISH} >_s \text{TOGAII} >_s \text{BOBCAT} \tag{1}$$

The relation $>_s$ means 'has a higher winning rate than' or 'is stronger than'. Here winning rate is calculated as

---

[1] http://wbec-ridderkerk.nl/html/UCIProtocol.html

Table 2. Enumeration of voting situations.

Situation	Leader's candidate move accepted/denied
All programs propose same move	Accepted
All programs propose different moves	Accepted
Leader is included in majority	Accepted
Leader is outvoted by majority	Denied

$$winning\_rate = \frac{number\_of\_wins + (0.5 * number\_of\_draws)}{number\_of\_games}$$

A round-robin tournament was played between STOCKFISH, TOGAII, and BOBCAT, in order to ensure condition 1 for each group. The tournament was played for $(8, 7, 9)$, $(9, 8, 10)$, $(10, 9, 11)$ and $(11, 10, 12)$. 1,000 games were played for each pair of programs. The winning rate of each pair was compared in order to calculate the relative strength. It was found that condition 1 was satisfied with 95 % significance using a one-tailed Binomial Test. However, the group $(8, 7, 9)$ did not satisfy the condition and so this group is not included in the results. $(8, 7, 8)$ did satisfy condition 1. Since we assume that BOBCAT with a smaller depth-limit will perform worse, thus, we assume that any groups made using BOBCAT with a smaller depth-limit than the one which has been statistically proven to match condition 1, will therefore also match condition 1. Thus, 27 of the 28 groups satisfied condition 1.

2.3 Experiments

The groups were matched against an individual program. STOCKFISH was used as the opponent, and was set to search to the same depth as the group leader STOCKFISH. For example, the Chess match $(D_S = 9, D_T = 8, D_B = 6)$ vs. $D_S = 9$. 1,000 games were played for each of the 27 matches. During the same experiments, statistics were gathered about the situations where the group made a collective decision which outvoted the group leader and denied the proposed candidate move of the leader program. See Table 2.

3 Results

Figure 1 shows the changing depth of BOBCAT against winning rate of the group. The figure also shows the 95 % confidence threshold calculated using a one-tailed Binomial Test (dashed line). As can be seen in this figure, simple majority voting using three separate Chess programs can beat the leader of the group. Strong groups can be seen on the right hand side of the graph, where the group is stronger than the leader.

We add this result to existing results in 3-member simple majority voting in games as seen in Table 3. Also, [8] applied simple majority voting to the game of

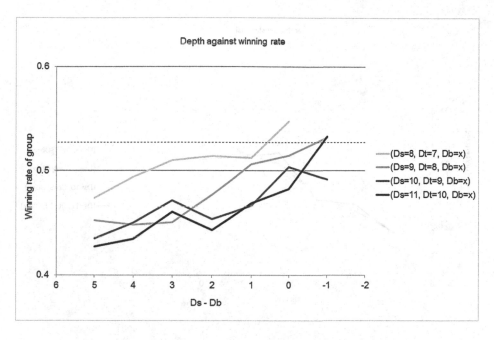

Fig. 1. Winning rate of different groups vs. Stockfish.

Table 3. Summary of results in applying 3-member simple majority voting algorithm to games.

Game	Group members	No. games	Relative strengths	Result
Chess [3]	Roma, Mach, MM	20	$R >_s M >_s MM$	$R >_s SMV(3) >_s MM$
Shogi [9]	Bonanza, YSS, GPS Shogi	1,000	$B >_s Y \approx_s G$	$SMV(3) >_s B$
Chess	Stockfish, TogaII, Bobcat	1,000	$S >_s T >_s B$	$SMV(3) >_s S$

Chess, however, in that work the group of programs were larger than 3 members, and consisted of one Chess program with a random modification. Our results on Chess are added to the bottom row of the table. Althöfer's results published in [3] do not have statistical significance which required 20 data points. Also, they were tested using a time limited search of 30 seconds per move. Obata *et al.* [9] did not include draws in the winning rate calculation for Shogi, and also the search limit was node-based. These differences between results make it difficult to compare them directly, but they are presented for reference.

Figure 1 also shows that increasing the strength of the weakest group member results in a stronger group when playing against the group leader. This is true for each of the four sets of groups.

Additionally, Fig. 1 shows that a group composed of weaker programs is a stronger group, when playing against the leader. This does not confirm the relative strength of different groups because the opponent's strength also varies.

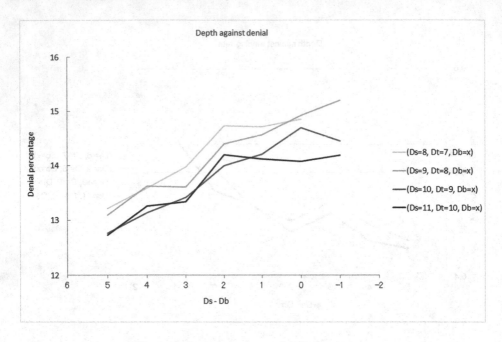

Fig. 2. Denial percentage of different groups.

Figure 2 shows the changing depth of BOBCAT against the denial percentage, where

$$denial\_percentage = \frac{denials}{group\_moves - opening\_book\_moves}$$

Here *denials* is the total number of denial situations in all games played by that group (see Table 2), *group_moves* is the total number of moves played by the group, and *opening_book_moves* is the total number of moves played from the opening book for that group.

As can be seen in Fig. 2, increasing the strength of the weakest group member results in an increasing probability of denial of the group leaders proposed action. This means that TOGAII and BOBCAT become increasingly more likely to agree on an alternative to the leader's proposal.

Also from Fig. 2 it can be seen that a group composed of weaker members, S_1 compared with S_4, is characterised by a larger percentage of denials. It is not clear why this is the case.

Figure 3 shows the relationship between the amount of leader denials and the strength of the group, and also a linear approximate curve is plotted. Two observations can be made. First, that within a given set, e.g., S_1, the strength of the group increases proportionally with the increasing probability of denying the group leader's proposed action. It is possible that there is a peak point, up to which the strength of group will increase, and then after which, as the denial of group leader increases towards 100 %, the strength of the group will decrease. The second observation from Figure 3 is that as the strength of the team members

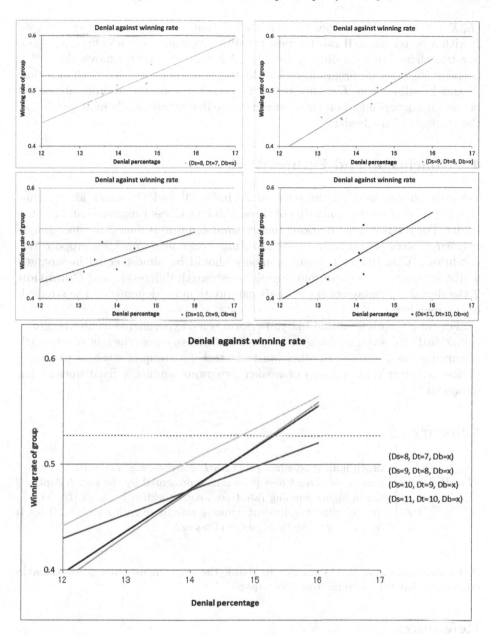

Fig. 3. Relationship of denial with winning rate.

increases, i.e., S_1 to S_4, then the denial percentage of a significantly stronger team decreases.

A theory can be started from these two observations. With 5 moves, m_1 to m_5, with $m_1 >_b m_2 >_b m_3 >_b m_4 >_b m_5$, where $>_b$ means 'better', and team

(L, X, Y). If L picks m_3 then there is a 50 % chance that X and Y will outvote L with a better move. If L is the best possible program, and picks m_1 then there is a 0 % chance that an outvote by X and Y will be a better move. Also, if L is the worst possible program, and picks m_5 then there is a 100 % chance that X and Y will outvote L with a better move. Therefore, the appropriate denial percentage is dependent on the strength of the individual members, in particular the strength of the leader.

4   Conclusion and Future Work

We have shown, as a continuation of Althöfer [3] and Obata *et al.* [9], that simple majority voting consisting of three different Chess programs can beat the leader. For the research question, *"under what conditions does 3-member simple majority voting play a better game than the leader program?"* we propose two conditions. **Condition 1** group members should be almost equal in strength whilst still showing a small, but significant strength difference, and **Condition 2** the denial percentage of the leader's candidate move depends on the strength of the members.

For future work we would like to perform some experiments where the group is matched against an unchanging program in order to assess the relative strength of varying sets of groups. In other words, to test if a group of stronger programs performs better than a group of weaker programs against a fixed unchanging opponent.

Glossary

D_P	depth limit used when program P searches, e.g. $D_S = 10$
(D_x, D_y, D_z)	a group of three Chess programs represented by the search depths
$>_s$	'has a higher winning rate than', e.g. Stockfish $>_s$ Toga II
$\approx_s$	'has an almost equivalent winning rate to', e.g. Stockfish $\approx_s$ Toga II
$>_b$	'is a better move than', e.g. m1 $>_b$ m2

Acknowledgments. We would like to thank the anonymous referees for valuable comments that helped to improve this paper.

References

1. Cook, D.: A human-computer team experiment for 9x9 go. In: van den Herik, H.J., Iida, H., Plaat, A. (eds.) CG 2010. LNCS, vol. 6515, pp. 145–155. Springer, Heidelberg (2011)
2. Shaw, M.E.: Comparison of individuals and small groups in the rational solution of complex problems. Am. J. Psychol. **44**, 491–504 (1932)
3. Althöfer, I.: Selective trees and majority systems: two experiments with commercial chess computers. In: Advances in Computer Chess 6, pp. 37–59 (1991)

4. Althöfer, I.: Improved game play by multiple computer hints. Theor. Comput. Sci. **313**, 315–324 (2004)
5. Hanawa M., Ito, T.: The optimal consultation system in a thought algorithm. In: The 3rd Symposium of Entertainment and Cognitive Science, pp. 72–75 (2009) (in Japanese)
6. Littlestone, N., Warmuth, M.: The weighted majority algorithm. In: Annual Symposium on Foundations of Computer Science (FOCS 1989), pp. 256–261 (1989)
7. Triplett, N.: The dynamogenic factors in pacemaking and competition. Am. J. Psychol. **9**, 507–533 (1898)
8. Omori S., Hoki K., Ito, T.: Consultation algorithm for computer chess. Technical report 2011-GI-26(5), pp. 1–7 (2011) (in Japanese)
9. Obata, T., Sugiyama, T., Hoki, K., Ito, T.: Consultation algorithm for computer shogi: move decisions by majority. In: van den Herik, H.J., Iida, H., Plaat, A. (eds.) CG 2010. LNCS, vol. 6515, pp. 156–165. Springer, Heidelberg (2011)
10. Sato Y., Cincotti A., Iida H.: An analysis of voting algorithm in games. In: Computer Games Workshop at European Conference on Artificial Intelligence (ECAI) 2012, pp. 102–113 (2012)

Comparison Training of Shogi Evaluation Functions with Self-Generated Training Positions and Moves

Akira Ura[1]([⊠]), Makoto Miwa[2], Yoshimasa Tsuruoka[1], and Takashi Chikayama[1]

[1] Graduate School of Engineering, The University of Tokyo,
Tokyo 113-8656, Japan
`ura@logos.t.u-tokyo.ac.jp`
[2] School of Computer Science, The University of Manchester,
Manchester M1 7DN, UK

Abstract. Automated tuning of parameters in computer game playing is an important technique for building strong computer programs. *Comparison training* is a supervised learning method for tuning the parameters of an evaluation function. It has proven to be effective in the game of Chess and Shogi. The training method requires a large number of training positions and moves extracted from game records of human experts; however, the number of such game records is limited. In this paper, we propose a practical approach to create additional training data for comparison training by using the program itself. We investigate three methods for generating additional positions and moves. Then we evaluate them using a Shogi program. Experimental results show that the self-generated training data can improve the playing strength of the program.

1 Introduction

Strong computer game programs for Chess-like games need evaluation functions that can accurately estimate the chance of winning for a given position. Many studies have been conducted to find effective ways of tuning the parameters of an evaluation function in an automatic fashion [10,15]. Among them, *comparison training* [18] has proven to be effective in the games of Chess and Shogi [11,13]. Comparison training is a supervised learning method that regards the moves played by human experts as the "correct" moves and tunes the parameters of the evaluation function so that the program makes the correct moves as often as possible. As with other supervised approaches, the effectiveness of this training method is, in large part, determined by the number of training positions. However, the number of game records of experts is usually limited, which motivated us to explore an approach to increase the number of training positions by the program itself.

Computer game programs have already become stronger than top human players in several games including Backgammon [19], Othello [5], and Chess,

H.J. van den Herik et al. (Eds.): CG 2013, LNCS 8427, pp. 208–220, 2014.
DOI: 10.1007/978-3-319-09165-5_18, © Springer International Publishing Switzerland 2014

as epitomized by the well-known victory by DEEP BLUE, IBM's Chess machine, over the then world Chess champion Garry Kasparov in 1997 [6]. For more complicated games, such as Shogi and Go, computer programs are not as strong as top human players but are strong enough to defeat top amateurs. In these games, moves made by computer programs with deep searches can be as good as moves by experts. Hence, our assumption is that moves and positions generated by computer programs can be used as additional training data for comparison training. One potential problem with this approach is the computational cost required to create such training data—it may take thousands of hours to perform deep searches for, for example, one million positions on a single machine. However, the process of generating such training data can be easily parallelized and we can solve the problem of computational cost by using a computer cluster (or a cloud service such as Amazon EC2).

In this paper, we investigate three methods for generating additional training data for comparison training and evaluate their effectiveness using GEKISASHI, a state-of-the-art Shogi program. The three generation methods correspond to three types of situations in which the evaluation function is used, namely, self-play games, game-tree search, and random games. We then evaluate the methods by measuring the playing strengths of the resulting evaluation functions.

This paper is organized as follows. Section 2 describes previous work on automated tuning of evaluation functions and methods that use search results of computer programs for tuning. Section 3 presents our proposed methods for generating training positions and moves and how to use them for training. Experimental results obtained by using a Shogi program are described in Sect. 4. Finally, Sect. 5 summarizes our contributions and describes some potential future work.

2 Related Work

There is a large body of previous work on automated tuning of parameters in computer game programs [10, 15, 20]. In particular, automated tuning of evaluation functions has been studied extensively [5, 11, 13, 14, 18, 21], and several methods have been proposed to conduct supervised learning using game records of human experts [11, 13, 18, 21]. In comparison training, the parameters of an evaluation function are adjusted in such a way that the mini-max value of the expert's move becomes higher (in the nega-max sense) than those of the other legal moves. This method was implemented in DEEP BLUE in a slightly modified form [18]. In Shogi, Hoki used a similar method to build an evaluation function for his program, BONANZA, and won the World Computer Shogi Championship [13]. This was the first successful application of automated tuning of evaluation functions in Shogi. Today, nearly all top Shogi programs employ a comparison training method (or its variant) to tune their evaluation functions.

In comparison training, the quality of the moves in the game records is especially important. That is, the moves in the training data should be sufficiently reliable that they can be regarded as correct moves. Kaneko investigated the influence of the quality of the game records on the resulting players using Shogi programs [12]. He compared three evaluation functions trained on three sets of game records that differed in quality. More specifically, the game records used were those of experts, amateurs, and computer players, each consisting of 10,000 games. The game records of the computer players were retrieved from *floodgate*[1]. They were generated by four strong computer programs in 30-min games. His experimental results showed that the player trained with the game records of experts was the strongest of the three. What is interesting is that the player trained with the game records of computer players was stronger than the one trained with those of amateurs.

There are many approaches to automated tuning of evaluation functions other than comparison training. Here, we focus on methods that use search results of computer players. Lee and Mahajan proposed a method to tune evaluation functions for Othello from data generated by self-play [14]. They used positions generated by playing 20 moves randomly after the initial position. Buro extracted positions that were evaluated in game-tree search and assigned scores to them by performing additional search [4]. The self-play approach has been widely used in the context of reinforcement learning [17], and was applied to Backgammon [19], Chess [1,22], and Shogi [2]. In evolutionary computation for evaluation functions, self-play games were performed to evaluate the fitness of individuals [3,9]. Tabibi et al. proposed an approach for building a computer player that mimics another player [8]. They assigned mini-max values to positions of experts using a reference computer game player, and tuned the parameters of an evaluation function using a genetic algorithm so that the function can produce values that are as close as possible to the assigned values for the positions.

3   Method

In this work, we propose three methods, which we call *Self-play*, *Leaf*, and *Random*, to create additional training data for comparison training by using the program itself.

The first two methods are developed in the hope that they would generate training positions that are similar to those arising in actual games. In principle, we should generate game positions worth being used for training, and it is important to use game positions that are similar to those arising in actual games since the evaluation functions are called to evaluate such positions and the weights of the features need to be appropriately tuned for them. A negative example of this would be generating positions where all pieces are randomly placed. These positions are most likely useless because they will not appear in actual games and share few features with the positions arising in actual games. The two methods are described more specifically as follows.

Self-play. Positions that appear in self-play of a computer player are used as training positions. These positions are expected to appear also in actual

---

[1] http://wdoor.c.u-tokyo.ac.jp/shogi/floodgate.html

games. We used positions in game records of experts as the initial positions for self-play to ensure variety of training positions.

Leaf. A game-tree search is performed for positions appearing in game records of experts and a fraction of the leaf positions in the search trees are used as the training positions. The idea of using such leaf positions has been proposed previously by Buro in a context different from comparison training [4]. This method is based on the idea that the training positions should share the same characteristics as the positions for which evaluation functions are called in actual game-tree searches.

The third method is presented mainly to provide a baseline. Unlike the first two methods, most of the positions generated by this method may not have characteristics similar to the ones appearing in actual games.

Random. Positions created by playing two legal moves randomly from positions in game records of experts are used as training data.

After the generation of the positions, the "correct" move is computed for each of the generated positions by performing a moderately long-time search using the program itself. The resulting positions and moves are then used for comparison training.

3.1 Comparison Training in Gekisashi

We used a computer Shogi program, GEKISASHI [20], throughout the experiments in this work. GEKISASHI is one of the state-of-the-art computer Shogi players, and it has won the World Computer Shogi Championship four times.

GEKISASHI uses a linear evaluation function:

$$f(\boldsymbol{w}, s) = \boldsymbol{w}^T \boldsymbol{\phi}(s) , \tag{1}$$

where $\boldsymbol{w}$ is a weight vector (parameters) and $\boldsymbol{\phi}(s)$ is a feature vector that represents a game position s. The total number of elements in the weight vector is approximately 1,900,000. The comparison training in GEKISASHI is performed by using an online training method called Averaged Perceptron [7]. First, the weights for pieces are initialized to some heuristic values and the other weights are set to zero. The perceptron update rule for the weight vector is as follows:

$$\boldsymbol{w}^{(t)} = \boldsymbol{w}^{(t-1)} + \frac{1}{|S^{(t)}|} \sum_{s_i \in S^{(t)}} (\boldsymbol{\phi}(s_1) - \boldsymbol{\phi}(s_i)) , \tag{2}$$

where t is the number of updates, s_1 is the leaf position on the principal variation calculated by search from the position after the expert's move, s_i $(i \geq 2)$ is the leaf position on the principal variation after a move the expert did not make, and $S^{(t)}$ is a group of leaf positions on principal variations after the moves which are evaluated to be better than the expert's move at the t-th update.

$$S^{(t)} = \{s_i | i \geq 2 \wedge f(\boldsymbol{w}^{(t-1)}, s_1) < f(\boldsymbol{w}^{(t-1)}, s_i)\} . \tag{3}$$

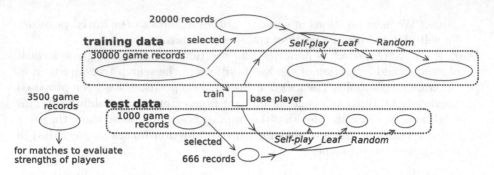

Fig. 1. Overview of how training data are generated.

The number of positions included in $S^{(t)}$ is denoted as $|S^{(t)}|$. In all training processes in this study, the total search depth is six, that is, the positions realized by each legal move are searched with a depth of five. The final weight vector is computed as the average of all intermediate weight vectors during the training:

$$\boldsymbol{w}^* = \frac{1}{T+1} \sum_{t=0}^{T} \boldsymbol{w}^{(t)} \ , \tag{4}$$

where T is the total number of updates in training.

In addition to training data, we also prepare test data and use this for the decision to terminate the training. Every position in the training data is seen once per iteration in training. After each iteration, we compute an average weight vector and count the number of "correct" moves that the weight vector makes on positions in the test data. The training is continued until the number starts to decrease. The final weight vector is the averaged weight vector that can make the largest number of "correct" moves among the averaged weight vectors computed by the iterations.

3.2   Data Generation and Training

Figure 1 shows an overview of how the training positions and moves are generated. First, we train a computer player by the training method described in Sect. 3.1 using game records of experts. We call this player *base player*. Next, we generate three kinds of new training and test data (*Self-play*, *Leaf*, and *Random*) using the base player. Lastly, we train two different players using each kind of generated data. One player is trained using the generated training data and game records of experts. This player is used to evaluate the data generation method. The other player is trained using only the generated training data. This player is used to evaluate the effect of the generated training positions in detail.

To generate *Self-play* positions, a self-play game is started from a position obtained by making the first 30 moves[2] from the initial position. Positions that appear in the self-play are added to the *Self-play* data. To avoid using duplicated positions, we remove positions that appear in the game record on which the self-play is started or have already appeared in the self-play game.

To generate *Leaf* positions from game records of experts, we perform a game-tree search on every position obtained by making more than 35 moves from the initial position by using the base player with a search depth of 14. The reason why we use this search depth is that the search depth in the training is six and the desirable search depth for real matches is at least $20 \ (= 14 + 6)$[3] in each game record of experts. After the search, we select a position randomly from the leaf positions in each game tree and add it to the *Leaf* data.

To generate *Random* positions from game records of experts, two random moves are made from every position obtained by making more than 35 moves from the initial position. If at least one of the two moves is different from the moves in the record, the position is added to the *Random* data.

4   Evaluation

In this section we start by describing the experimental settings (Sect. 4.1). then we evaluate the strengths of the players using both 30,000 game records of experts and self-generated training data (Sect. 4.2). We conclude the section by analysing the effects of the different training data (Sect. 4.3).

4.1   Experimental Settings

Our experimental environments consisted of Xeon E5530, Xeon E5620, Xeon X5687, and Xeon E5-2665 (1,688 cores in total). We used 30,000 game records for training and 1,000 game records for testing (Fig. 1). For each data generation method described in Sect. 3, we generated training positions from 20,000 game records and test positions from 666 game records. The training positions were searched by the base player in order to obtain the training moves that are used as the correct moves in the comparison training. The number of nodes searched was 30,000,000, 3,000,000, or 300,000. It took about 120 s to search 30,000,000 nodes on one core of Xeon E5530 (2.40 GHz). We removed the training positions where either of the kings was found to be checkmated during the search. Table 1 shows the number of training positions for each kind of training data when 30,000,000 nodes were searched to generate the moves.

---

[2] We can get a sufficient variety of game positions by making the first 36 moves from game records of experts. A shogi game is usually still in the opening stage even after playing the first 36 moves. The generation of *Leaf* and *Random* is done with 35 moves while *Self-play* uses only 30 moves because the base player may make the same moves as experts. Some extra moves of the base player are needed in *Self-play* to generate different positions from game records of experts.

[3] It takes several tens of seconds for GEKISASHI to perform a search with a depth of 20 in a typical middle-game position.

Table 1. The number of positions in training data

Training data	# of positions
Experts' 30,000 game records	2,859,698
Self-play from 20,000 game records	1,862,518
Leaf from 20,000 game records	1,287,870
Random from 20,000 game records	1,381,263

We set aside 3,500 game records of experts in order to evaluate the playing strengths of the resulting players. These game records were used to create the initial positions for matches between any two players. Each initial position was created by selecting the position just after the first 36 moves of each game record. For each initial position, two matches were conducted by swapping black and white, and thus 7,000 matches were played in total. A match was considered a draw when the match did not finish in 200 moves. The number of nodes searched for each move in the matches was limited to 100,000 nodes.

4.2 Evaluation for Players Using both Game Records of Experts and Self-Generated Training Data

We evaluated the strengths of the players using both the 30,000 game records of experts and the generated training data against two opponent players. We used searches with 30,000,000 nodes (about 120 s) to generate the training and test data. The test data consisted of the 1,000 game records of experts and the same kinds of the generated test data as the generated training data.[4] Between each pair of players, 7,000 games were played (see Sect. 4.1). For the opponent players, we used the base player that was trained with the 30,000 game records and a player that was trained with 40,000 game records of experts. The former player was used to confirm whether the additional training data could increase the strength of the base player. The latter player was prepared to find out whether or not players using the additional training data were as strong as the player using additional 10,000 game records of experts.

The results are shown in Table 2. The result of games between the player using the 40,000 game records ("10,000 game records" in the table) and the base player is also shown for comparison. The two players using the *Leaf* data (*Leaf* and *Self-play* + *Leaf*) were stronger than the base player and these two players did not show any significant difference in strength against the player using the 40,000 game records, while other players without the *Leaf* data did not show any significant difference against the base player and some of the players were significantly weaker than the player using the 40,000 game records. These results indicate that the *Leaf* method is the most effective in the three methods to generate training positions. As additional experiments, we generated *Leaf* data

---

[4] For example, when the training data included the *Leaf* training data and the *Random* training data, the test data included the *Leaf* test data and the *Random* test data.

Table 2. Winning rates of the players that use both 30,000 game records of experts and additional training data against two opponent players. The number of nodes searched to generate "correct" moves was 30,000,000 (about 120 s). Asterisks indicate statistical significance levels as to whether their strengths are different: * for $p < 0.05$, ** for $p < 0.01$, and *** for $p < 0.001$. The theoretical value (50.00 %) is shown when a player plays against itself.

Additional training data	vs 30,000 records (%)	vs 40,000 records (%)
10,000 game records	52.76***	50.00
(40,000 records in total)		
Self-play	49.92	47.12***
Leaf	51.51*	48.85
Random	49.04	48.33**
Self-play + *Leaf*	51.27*	48.91
Self-play + *Random*	50.56	48.89
Leaf + *Random*	50.48	50.16
Self-play + *Leaf* + *Random*	50.50	50.29

Table 3. Winning rates of the players that use both 30,000 game records of experts and additional *Leaf* data against two opponent players. The winning rates of *Leaf* from 20,000 game records in Table 2 are also shown.

Additional training data	vs 30,000 records (%)	vs 40,000 records (%)
Leaf from 20,000 game records	51.51	48.85
Leaf from 30,000 game records	51.52	49.68
Leaf from 40,000 game records	49.79	49.31
Leaf from 50,000 game records	50.49	49.01

using more than 20,000 game records and trained players. The results are shown in Table 3. Somewhat counterintuitively, however, an increase in *Leaf* data did not lead to any further improvement.

4.3   Analyses on the Effects of Different Training Data

To identify potential underlying causes of the experimental results presented in the previous section, we further analyze the effects of the training data from two different perspectives: the strengths of the players that generated the training data and the situations where they were generated.

We first examined how the strength of a player that generated training data affected the strength of a player that was trained on the generated training data. More specifically, we conducted experiments to investigate the following research questions.

– There are considerable differences in strength even among human experts. Is a player trained with game records of strong human experts stronger than a player trained with those of weak human experts?

– Computer players become stronger as their search becomes deeper. Are training data generated by deep searches more useful than training data generated
by shallow searches?
– Computer Shogi players may be as strong as human experts. Is a player using
training data generated by a computer player alone as strong as a player
trained with game records of experts?

For the evaluation, we prepared the following training data.

Experts. Positions after the first 36 moves in each game record of experts were
selected. Moves played by the experts were used.

High-rating experts (High-rating). Positions and moves played by high
-rating experts were selected from the positions in *Experts*. The ratings of the
experts were estimated from their game records by using a simplified variant
of the Elo rating system. 52 % of the positions in *Experts* were selected.

Low-rating experts (Low-rating). Positions and moves played by low-rating
experts were selected from the positions in *Experts*. 49 % of the positions in
Experts were selected.

Base player. Positions were the same as *Experts*, but moves were generated
by searches of the base player.

We trained players using 1,200,000 positions selected from each kind of training
data. For *Base player*, *Self-play*, *Leaf*, and *Random*, we used three different numbers of searched nodes (in Sect. 4.1) to generate the training data. We prepared
test data from the positions and moves of *Experts*. We chose two opponent players to evaluate the playing strengths. One was a player trained by the *Experts*
training data. The other was a player trained by the *Leaf* training data generated
by searches with 30,000,000 nodes because the player did not use game records
of experts and the *Leaf* training data were effective in the previous experiment
(in Table 2).

The results are shown in Tables 4 and 5. They give the following answers
to the aforementioned questions, and the answers are not dependent on the
opponent players.

– The players using the *High-rating* training data were stronger than the players
using the *Low-rating* training data (see Table 4). This result was consistent
with the result reported by Kaneko [12].
– The players using the training data generated by deep searches were, in most
cases, stronger than the players using the training data generated by shallow
searches (see Table 5). The player using *Base player* (30,000,000) was the only
exception.
– The players using game records of experts were significantly stronger than
the ones using self-generated training data (see Tables 4 and 5). This is partly
because players searching 30,000,000 nodes were not as strong as human
experts.

Next, we analyzed the effects of the situations where the training data were
generated (see Table 5). The results show that the *Random* data were less effective than the *Base player*, *Self-play*, and *Leaf* data. The results also show that

Table 4. Winning rates of the players that use moves played by human experts against two opponent players (* for $p < 0.05$, ** for $p < 0.01$, and *** for $p < 0.001$).

Training data	vs *Experts* (%)	vs *Leaf* (30,000,000) (%)
Experts	50.00	55.40***
High-rating	50.91	56.72***
Low-rating	48.42*	54.87***

Table 5. Winning rates of the players that use moves generated by the base player against two opponent players. The numbers of nodes searched to obtain the "correct" moves for comparison training are shown in the parentheses.

Training data	vs *Experts* (%)	vs *Leaf* (30,000,000) (%)
Base player (300,000)	40.43	46.29
Base player (3,000,000)	42.52	48.56
Base player (30,000,000)	41.78	47.75
Self-play (300,000)	41.85	47.68
Self-play (3,000,000)	42.82	48.42
Self-play (30,000,000)	43.69	49.58
Leaf (300,000)	41.92	46.66
Leaf (3,000,000)	42.23	49.05
Leaf (30,000,000)	44.60	50.00
Random (300,000)	40.03	45.27
Random (3,000,000)	39.38	45.50
Random (30,000,000)	40.63	46.20

positions in the *Self-play* and *Leaf* data were equally useful. These results differ from the results in Table 2, which used the generated data as additional data to the game records of experts in training. A possible explanation for the difference is that the positions in the *Leaf* data were different from those in the game records and had a positive influence on the training, while the positions in the *Self-play* data were too similar to positions in the game records.

The weakness of the players trained on the generated data may be attributed to "self-generation". To isolate the influence of self-play, we retrieved game records of high-rated computer shogi players[5] except GEKISASHI from floodgate and trained a player as well as the players described in this section. We used only the moves that expended more than five seconds (15 s on average). The number of training positions was 1,200,000. Winning rates of the resulting player were 44.51 % against *Experts* and 50.62 % against *Leaf* (30,000,000), while the winning rates of *Self-play* (30,000,000 nodes; 120 s) were 43.69 % and 49.58 % respectively (see Table 5). This shows that self-generation may have a bad effect on training.

---

[5] Players with a rating higher than 2550 as of June 10, 2013.

5   Conclusion and Future Work

We proposed an approach to generating training data by deep searches of a computer game player and to use them as additional training data to game records of experts in order to address the problem in comparison training that the number of the game records is limited. We evaluated three methods to generate training positions that differ in situations where the positions appeared. Our experimental results show that the method which extracts leaf positions from search trees was effective in generating training data when we used the training data as additional data to game records of experts. This is presumably because these leaf positions have different characteristics from those in experts' game records and they are similar to the ones that appear in actual game-tree searches.

Furthermore, we analyzed the effects of the training data from two perspectives: the strength of a player that generated the data and the situations where the data were generated. The players using training data generated by deep searches were stronger than the ones using training data generated by shallow searches. However, the players using the self-generated training data were significantly weaker than the players using game records of experts even when the search time to compute "correct" moves was set to about 120 s (30,000,000 nodes). The search time may have not been long enough to obtain moves that are as reliable as moves of experts. In terms of the situations where the data were generated, we confirmed that positions generated by playing moves randomly were not useful.

The experimental results suggest that the selection of the training positions significantly affects the strengths of the resulting players. One direction of future work is to explore more effective selection methods to improve the strength of players. One potential approach would be to select leaf positions that could change principal variations with slight changes in their values, since the accurate evaluation of these positions are required to select correct principal variations. A second potential approach would be to use win/loss information. Sato et al. proposed a method introducing not only the agreement of the moves but also the win/loss information to the objective function [16]. They reported that the win/loss information was important for some game positions. This method can be also effective when self-generated positions and moves are used.

A new direction of future work is to use several other programs along with GEKISASHI to generate training data. Baxter et al. reported that their Chess program became stronger using reinforcement learning when it played against many players including human players on a Chess server [1]. We also observed that the self-generated data may have a bad effect on training. These results suggest that we could build a stronger player by using training data generated by several computer programs.

References

1. Baxter, J., Tridgell, A., Weaver, L.: Reinforcement learning and chess. In: Furnkranz, J., Kubat, M. (eds.) Machines That Learn to Play Games, pp. 91–116. Nova Science Publishers, Inc., Hauppauge (2001)
2. Beal, D.F., Smith, M.C.: Temporal difference learning applied to game playing and the results of application to shogi. Theor. Comput. Sci. **252**(1–2), 105–119 (2001)
3. Bošković, B., Brest, J., Zamuda, A., Greiner, S., Žumer, V.: History mechanism supported differential evolution for chess evaluation function tuning. Soft Comput. **15**(4), 667–683 (2010)
4. Buro, M.: From simple features to sophisticated evaluation functions. In: van den Herik, H.J., Iida, H. (eds.) CG 1998. LNCS, vol. 1558, pp. 126–145. Springer, Heidelberg (1999)
5. Buro, M.: Improving heuristic mini-max search by supervised learning. Artif. Intell. **134**(1–2), 85–99 (2002)
6. Campbell, M., Hoane, A., et al.: Deep blue. Artif. Intell. **134**(1–2), 57–83 (2002)
7. Collins, M.: Discriminative training methods for hidden Markov models: theory and experiments with perceptron algorithms. In: EMNLP '02, pp. 1–8. Association for Computational Linguistics (2002)
8. David-Tabibi, O., Koppel, M., Netanyahu, N.S.: Expert-driven genetic algorithms for simulating evaluation functions. Genet. Program. Evolvable Mach. **12**(1), 5–22 (2011)
9. Fogel, D.B., Hays, T.J., Hahn, S.L., Quon, J.: A self-learning evolutionary chess program. Proc. IEEE **92**(12), 1947–1954 (2004)
10. Fürnkranz, J.: Machine learning in games: a survey. In: Fürnkranz, J., Kubat, M. (eds.) Machines That Learn to Play Games, pp. 11–59. Nova Science Publishers, Inc., Hauppauge (2001)
11. Hoki, K., Kaneko, T.: The global landscape of objective functions for the optimization of shogi piece values with a game-tree search. In: van den Herik, H.J., Plaat, A. (eds.) ACG 2011. LNCS, vol. 7168, pp. 184–195. Springer, Heidelberg (2012)
12. Kaneko, T.: Evaluation functions of computer shogi programs and supervised learning using game records. J. Jpn. Soc. Artif. Intell. **27**(1), 75–82 (2012) (In Japanese)
13. Kaneko, T., Hoki, K.: Analysis of evaluation-function learning by comparison of sibling nodes. In: van den Herik, H.J., Plaat, A. (eds.) ACG 2011. LNCS, vol. 7168, pp. 158–169. Springer, Heidelberg (2012)
14. Lee, K.F., Mahajan, S.: A pattern classification approach to evaluation function learning. Artif. Intell. **36**(1), 1–25 (1988)
15. Mandziuk, J.: Knowledge-Free and Learning-Based Methods in Intelligent Game Playing. Springer, Heidelberg (2010)
16. Sato, Y., Miwa, M., Takeuchi, S., Takahashi, D.: Optimizing objective function parameters for strength in computer game-playing. In: AAAI '13, pp. 869–875 (2013)
17. Sutton, R.S., Barto, A.G.: Reinforcement Learning: An Introduction. Cambridge University Press, Cambridge (1998)
18. Tesauro, G.: Comparison training of chess evaluation functions. Machines That Learn To play Games, pp. 117–130. Nova Science Publishers, Inc., New York (2001)
19. Tesauro, G.: Programming backgammon using self-teaching neural nets. Artif. Intell. **134**(1–2), 181–199 (2002)
20. Tsuruoka, Y., Yokoyama, D., Chikayama, T.: Game-tree search algorithm based on realization probability. ICGA J. **25**(3), 146–153 (2002)

21. Vázquez-Fernández, E., Coello, C.A.C., Troncoso, F.D.S.: An evolutionary algorithm coupled with the Hooke-Jeeves algorithm for tuning a chess evaluation function. In: IEEE CEC '12, pp. 1–8 (2012)
22. Veness, J., Silver, D., Uther, W., Blair, A.: Bootstrapping from game tree search. Adv. Neural Inf. Process. Syst. **22**, 1937–1945 (2009)

Automatic Generation
of Opening Books for Dark Chess

Bo-Nian Chen and Tsan-sheng Hsu$^{(\boxtimes)}$

Institute of Information Science, Academia Sinica, Taipei, Taiwan
{brain,tshsu}@iis.sinica.edu.tw

Abstract. Playing the opening game of dark chess well is a challenge that depends to a large extent on probability. There are no known studies or published results for opening games, although automatic generation of opening books for many games is a popular research topic. Some researchers collect masters' games to compile an opening book; while others automatically collect computer-played games as their opening books. However, it is difficult to obtain a strong opening book via the above strategies because few games played by masters have been recorded. In this paper, we propose a policy-oriented search strategy to build automatically a selective opening book that is helpful in practical game playing. The constructed book provides positive feedback for computer programs that play dark chess.

1 Introduction

Dark chess is a two-player zero-sum probability game that is popular in Asia [1]. The game was established as a tournament event at the Computer Olympiad in 2010 (the 15th Computer Olympiad). Dark chess has a distinct property namely that all the pieces are "dark" (i.e., each piece's type is unknown) at the starting positions on the board. A player discovers a piece's type and color by flipping it. This means the positions' scores depend on the probabilities of revealing each type of piece. During a game, each player performs two actions: *flip* and *move*. There is also a special piece, the cannon, which captures pieces by jumping over a piece called a *carriage*. This characteristic is inherited from Chinese chess.

The opening game in dark chess relies primarily on effective flipping strategies. If the material revealed on the board gives one player an advantage, he[1] is likely to win the game. Strong pieces revealed during the opening game are dangerous because they have little mobility and can be easily captured by the other player performing a surprise flip move. How to protect important revealed pieces is also a critical problem in the opening game. In the middle game, moving pieces to important locations and capturing the opponent's strong pieces are crucial to winning the game; and in the endgame, a player needs to capture all opponent's pieces in order to win the game. A player's skill in moving his pieces to capture his opponent's pieces determines whether he will win the game.

---

[1] For brevity, we use 'he' and 'him' whenever 'he or she' and 'his or her' are meant.

H.J. van den Herik et al. (Eds.): CG 2013, LNCS 8427, pp. 221–232, 2014.
DOI: 10.1007/978-3-319-09165-5_19, © Springer International Publishing Switzerland 2014

To handle the opening game in dark chess well, the following three strategies may be applied: (1) cascade a strong computer program with an evaluation function, (2) compile an opening book manually or by collecting a large number of masters' games, or (3) perform self-play games with randomly selected first n moves. In the first method, the moves suggested by a search algorithm are deterministic. Books constructed by the second method may be affected by two challenges: (a) the reliability of the books cannot be guaranteed; and (b) a good move that was played frequently in the past may become a bad move in the future as more knowledge is discovered. The third method would need a very large number of games to obtain an opening book that could be used in tournaments. In summary, a method that can automatically generate effective opening moves and avoid a large search space is necessary. To this end, we propose a policy-oriented method for constructing a practical opening book for dark chess.

The remainder of this paper is organized as follows. In Sect. 2, we discuss state-of-the-art opening book construction approaches, utilization techniques, and critical issues. In Sect. 3, we define the opening policies and consider the selection of the policies; and in Sect. 4, we describe the proposed policy-oriented search technique. In Sect. 5, we evaluate the technique in experiments on different versions of the dark chess program FLIPPER. Section 6 contains some concluding remarks.

2   Previous Work

The opening game is usually approached by constructing an opening book. Some designers manually edit an opening book for tournaments based on their knowledge of the game or existing theory, while others create opening books by collecting a large number of masters' games [2]. In these types of books, the opponent can force the other player to drop out of the opening book by using one of the following two approaches. If the opponent is stronger, he might select a slightly disadvantageous variation of a move to force the other player to drop out and beat the program in the middle game or endgame. Second, he could select an uncommon variation to increase the probability that the other player will drop out. Automatic opening book generation schemes have been proposed to handle the above situation. Opening books are generated with the best-first search algorithm [2,3] or the Monte-Carlo tree search algorithm [4]. The former is applied in chess-like games; and the latter is applied in games such as Go.

The process for generating an opening book is similar to playing a game except that it is performed off-line. Fast computation is not necessary for propagation, but accurate evaluation of the positions is essential [5]. The content of an opening book can be divided into three types: (1) the position and move lists; (2) the game collection with statistical information [3]; and (3) the states of a game board [6,7]. The constructed opening book can be used as a dictionary [6,8], or as part of the evaluation function [9].

Because of the probabilistic nature of dark chess, the search space of an opening game is very large. In addition, according to researchers who have designed opening books for other games, "domain-dependent knowledge" is a critical technique for pruning an opening book. To avoid the problem of a very large state space and provide a flexible opening book for different players, we propose a policy-oriented method to derive a feasible opening book for dark chess.

3   Opening Policies

A *policy* is a set of moves designed to achieve a certain goal. Instead of studying moves, it might be easier to devise suitable opening strategies and find general rules for the strategies by focusing on policies.

Opening policies are divided into attack policies and non-attack policies. The former focus on capturing the opponent's pieces; while the latter deal with factors other than materials that are critical in the opening game.

3.1   Attack Policies

The first player selects a piece to flip. Then, the second player attacks the piece revealed by the first player. There are two kinds of attack strategies.

Adjacency attack: the player tries to reveal a piece adjacent to one of the opponent's pieces and to capture it.
Cannon attack: the player tries to reveal a cannon to capture one of the opponent's pieces.

For example, if the first player reveals a king, an adjacency attack would have a higher chance of success than a cannon attack because there are five pawns that can capture a king, but only two cannons can capture it. However, if a cannon is the target, only an adjacency attack would be effective.

The above strategies have different success rates when different pieces are attacked. Table 1 shows that after the first player reveals a piece, the second player launches an adjacency attack to capture the piece at ply 2. The term *success* means the second player reveals a piece that can capture the first player's piece, but the first player's piece cannot capture the second player's piece. Thus, a successful attack means the second player will be able to capture the first player's piece in his next turn. *Fatal* indicates that the revealed piece can be captured directly by the first player's piece. *Other* denotes that the two pieces cannot capture each other because (1) they are on the same side; (2) they are

Table 1. Success rates of adjacency attacks

Type	King	Guard	Minister	Rook	Knight	Cannon	Pawn
Success	5 (16 %)	1(3 %)	3 (10 %)	5 (16 %)	7 (23 %)	9 (29 %)	8 (26 %)
Fatal	11 (35 %)	15 (48 %)	13 (42 %)	11 (35 %)	9 (29 %)	0 (0 %)	6 (19 %)
Other	15 (48 %)	15 (48 %)	15 (48 %)	15 (48 %)	15 (48 %)	22 (71 %)	17(55 %)

both cannons; or (3) one of the pieces is a cannon and the other is a pawn. In Table 1, each element includes the number of pieces and the ratio of the number of pieces divided by a factor of 31. For example, if the second player flips a piece to capture a king at ply 2, there are 5 pieces that can achieve the goal, i.e., 5 black pawns. The success rate is 5/31 = 16 %.

Thus, the dark chess opening game can be modeled as a decision tree in which there are two types of nodes: *position nodes* for actual positions and *probabilistic nodes* for the positions yet to be played. The latter expand a set of position nodes, which are the possible results of a certain sequence of decisions or moves with a given probability. Figure 1 shows the decision tree for the position nodes after the first player flipped $b7(G)$, and the second player considers two attack strategies. Cannon attacks have a fixed success rate of 2 (6 %), a fatal rate of 0, and a high rate of 29 (94 %) to reveal a piece that does not attack any other piece. Note that revealing a piece for a cannon to capture involves different behavior. Its success rate is 14 (45 %); its fatal rate is 2 (6 %); and the probability of revealing a piece that does not attack any other piece is 15 (48 %).

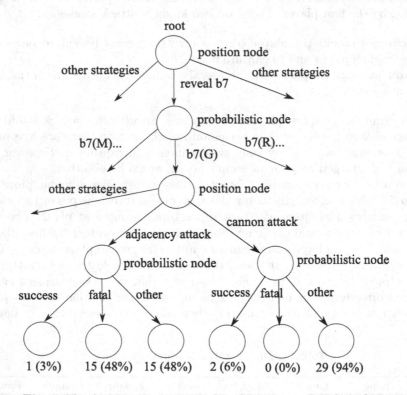

Fig. 1. The decision tree model for the opening game in dark chess

3.2   Non-attack Policies

An aggressive player may not always try to capture the opponent's pieces. By centralizing his pieces, a player can increase his advantage because pieces in a corner of the board are easier to control than those in the center. The player may also consider ways to improve the mobility of his pieces, or ways to protect his revealed pieces. If the opponent has established a strong claim to territory, finding a new region to develop may be a better choice than revealing pieces in the opponent's territory. We discuss these scenarios below.

Revealing Central Pieces. In general, pieces in the corner are likely to be controlled by central pieces and have very little mobility. This limits the power of the piece being controlled.

Defensive Moves. In an opening game, important pieces, such as kings, guards, ministers and cannons with zero mobility, can be captured easily by adjacency attacks or cannon attacks. Hence, another strategy involves protecting the revealed pieces under both adjacency and cannon attacks.

The results in Table 1 show that adjacency attacks on guards and ministers have a low success rate and a relatively high fatality rate. Therefore, to prevent adjacency attacks, we only need to protect kings and cannons. In cannon attacks, kings, guards, and ministers need to be protected. Cannons are not affected by cannon attacks.

Increasing Mobility. If a player flips an important piece, e.g., a guard, he needs to increase its mobility by revealing adjacent places. Obviously, the opponent will not reveal those places because of the risk of being captured.

New Territory. In dark chess, there is always a possibility that a player will maintain all his pieces on one part of the board, called his *territory*. Revealing pieces in the opponent's territory increases the probability that they will be captured. In such cases, revealing pieces in another part of the board to create a new territory may be a suitable choice.

In practice, we define a constant value $\delta = 500$ as the threshold that is applied to new territory when the opponent has an advantage score δ. The score $\delta = 500$ indicates that the difference of the position score is one guard or two ministers.

3.3   Policy Selections

Adjacency and cannon attacks are necessary and effective policies in an opening book. With regard to non-attack policies, some are suitable for use in the opening book of the first player, while others are suitable for the opening book of the second player. Thus, we design asymmetrical opening books for first and second players. The experiment results are detailed in Sect. 5.

4   Opening Book Generation and Queries

In this section, we discuss the proposed method for compiling an opening book that can be used to investigate the importance of the opening game and the initiative in dark chess.

4.1   Policy-Oriented Method

It is hard to find knowledge or books about opening strategies in dark chess. However, it is also difficult to obtain information about games played by experts or masters. Hence, we derive such knowledge by constructing an opening book. The only way to compile an opening book is to exploit an automatic generation method. To ensure that the book is not too predictable, we do not include self-play games in it. Moreover, we consider attack and non-attack policies when constructing an opening book. This avoids the search space to be too large and ensures that the book is effective.

Our automatic book generation algorithm includes system parameters that assign the policies needed during the generation process.

4.2   The Evaluation Function

In 2001, J. C. Chen proposed a method for compiling an opening book that can be stored by using hash table techniques [10]. The hash key in dark chess indicates the status of all places on the board, the number of each type of piece, and whether the red or black side will make the next move.

The evaluation function is essential for generating an opening book; however, it is currently difficult to design a good evaluation function that incorporates multiple facets of a game. We also believe that ordinal players care mainly about material gains in the opening game. Thus, we only consider the material values and make an adjustment for centralization to capture the positional values.

In the traditional max-min search algorithm, the return score indicates the path used to obtain that score. However, because a flipping action in dark chess does not produce a deterministic result, it is necessary to use a statistical method to estimate the result of a flipping action.

- $p.D$ denotes the set of dark pieces that have not been flipped yet.
- For a $d \in p.D$ and a strategy s

- $p_{s,d} = \text{get\_revealed\_position}(p, s, d)$
- $v_{s,d} = \text{max\_min}(p_{s,d})$ and $v_{s,d} \in V_{s,p.D}$
- Expected value $\mu = E(V_{s,p.D})$
- Standard deviation $\sigma = \sqrt{E(V_{s,p.D}^2) - [E(V_{s,p.D})]^2}$

The obtained μ provides an estimation of how good the revealing action is, but it is not comparable to the result computed by the evaluation function.

The latter is obtained by considering deterministic positions, such as moving actions or capturing actions [1].

A better strategy would be to include a standard deviation σ to obtain a bound with confidence levels v_l and v_r as the lower and upper bound respectively.

4.3 The Generation and Query Algorithm

To generate an opening book, we perform a max-min search, as shown in Algorithm 1. The off-line search yields more precise results and a more reliable opening book than applying a search algorithm directly.

Algorithm 1. Book generation algorithm

---

  procedure BookGen(position p)
  $strategy\_set = select\_strategy(p)$
  for all s in $strategy\_set$ **do**
    if s is a flipping action **then**
      $t = t_2 = 0$
      for all d in $p.D$ **do**
        $v_{s,d} = \text{max\_min}(get\_revealed\_position(p, s, d))$
        $t = t + v_{s,d}, t_2 = t_2 + v_{s,d}^2$
      $\mu = t/|p'.D|, \sigma = \sqrt{t_2 + t^2}$
      $p.v_l = \mu - \sigma \times r_d, p.v_r = \mu + \sigma \times r_d$
    else
      $v = \text{max\_min}(get\_move\_position(p, s))$
      $p.v_l = p.v_r = v$
    unmove()
    add_node(p)
  end procedure

In Algorithm 1, the $select\_strategy$ function is a user-defined method that allows designers to define their own opening policies according to a given position p. The algorithm performs a max-min search to retrieve the value of position p. If the move to be handled is a revealing action, the algorithm computes the expected value and the standard deviation of the search results of all possible pieces revealed by the move. Finally, we obtain $p.v_l$ and $p.v_r$ as the values of position p, and add p to the opening book. Two strategies can be used to compile an opening book. One strategy generates a very large book that can only be stored on a disk and retrieved by a random access technique; the other strategy tends to generate a smaller book that can be stored in the memory. The former includes much more data in the book; while the latter ensures the data can be accessed quickly. In practice, the smaller book is more feasible because human expert players seem to have only a limited amount of knowledge about the opening game, i.e., at most three branches for the majority of opening positions.

After Algorithm 1 completes a search, it stores the results as nodes or positions in a hash table. The strategy reduces the amount of memory needed for storage.

When a position is fed into the query algorithm (Algorithm 2) as a probabilistic node, the algorithm should retrieve all queriable successor moves from the book and compute estimated values for them. Note that users who design their own estimation functions can also apply them during the query procedure.

Algorithm 2. Book query algorithm

---

   function BookQuery(position p)

   $s_l = s_r = -\infty$

   action a = null

   move_gen(p, successor move set S)

   for all s in S **do**

     if s is a flipping action **then**

       $t = t_2 = 0$

       for all d in $p.D$ **do**

         $v_{s,d}$ = query(get_revealed_position(p, s, d))

         $t = t + v_{s,d}$, $t_2 = t_2 + v_{s,d}^2$

       $\mu = t/|p.D|$, $\sigma = \sqrt{t_2 + t^2}$

       $v_l = \mu - \sigma \times r_d$, $v_r = \mu + \sigma \times r_d$

     else

       v = query(get_move_position(p, s))

       $v_l = v_r = v$

     if $s_r < v_l$ **then**

       $s_l = v_l$, $s_r = v_r$, $a = s$

   return a

   end procedure

Actually, the BookQuery function can be integrated into the search engine. The search algorithm is implemented if no moves are retrieved. We define a ply threshold k. After k plies, the BookQuery algorithm is not called again.

5 Experiments

We performed three experiments to demonstrate the utility of policy-oriented method. Each node contains the value of a position computed by Algorithm 1 based on the evaluation function described in Sect. 4.2. If the move used in the current position is not in the opening book, the algorithm returns $-\infty$. The depth of the opening book in Algorithm 1 is set at 6 in this paper. The system parameter r_d used in Algorithm 2 is set at 1.5.

5.1 Experiment Design

The experiments are performed on the dark chess program called FLIPPER, which was developed by B. N. Chen and T.-s. Hsu [11]. Our baseline program is the original FLIPPER without an opening book. The FLIPPER_att version only uses attack policies as its opening book; while the FLIPPER_F version and the FLIP-PER_S version contain advanced policies for the first player and second player, respectively. The time settings in the experiments are 5 s for each move, and 20 min for a whole game. If a game goes to the same positions three times, or if neither side captures or reveals any pieces in 40 plies, the game is deemed a draw. Each set of experiments contains 200 games. Each side plays 100 games as the first player.

In the experiments, certain subsets of the policies are attached to each version. The results of first experiment, FLIPPER_att versus FLIPPER, demonstrate the effectiveness of the attack policies in the opening book. The second experiment includes the following versions:

1. FLIPPER_cen (reveal central pieces)
2. FLIPPER_def (defensive moves)
3. FLIPPER_mob (increasing mobility)
4. FLIPPER_ter (new territory)

The above versions exploit attack policies plus one of the non-attack policies. We try to find an acceptable combination of the policies to construct an opening book that is better than that of FLIPPER_att. In the second experiment, FLIPPER_att is used to play against the above four versions.

After the second experiment, we construct FLIPPER_F and FLIPPER_S. In the third experiment, we evaluate the performance of FLIPPER_F vs. FLIP-PER_att and FLIPPER_att vs. FLIPPER_S. The results demonstrate that combining advanced policies further improves the FLIPPER_att opening book.

5.2   Results and Discussion

We utilize the multi-nomial distribution model [12] to analyze the significance of an experimental result. Assume an experiment consists of $2n$ games, namely $g_1, g_2, ..., g_{2n}$. Let g_{2i-1} and g_{2i} be the ith pair of games. Let the outcome of the ith pair of games be a random variable X_i from the prospective of the copy who plays g_{2i-1}. We assign the value x for a game won, a score of 0 for a game drawn and a score of $-x$ for a game lost. The outcome of X_i and its occurrence probability is thus

$$Pr(X_i) = \begin{cases} p(1-p-q) & \text{if } X_i = 2x \\ pq + (1-p-q)q & \text{if } X_i = x \\ p^2 + (1-p-q)^2 + q^2 & \text{if } X_i = 0 \\ pq + (1-p-q)q & \text{if } X_i = -x \\ (1-p-q)p & \text{if } X_i = -2x \end{cases}$$

The mean $E(X_i) = 0$ and the standard deviation of X_i is

$$\sqrt{E(X_i^2)} = x\sqrt{2pq + (2q + 8p)(1 - p - q)}.$$

X_i is a multi-nominally distributed random variable. Let $X[n] = \sum_{i=1}^{n} X_i$. After playing n pairs of games, the probability of getting a score s is $Pr(|X[n]| = s)$. If $Pr(|X[n]| < s)$ is close to 1, the experimental result s can be deemed significant because with a probability $1 - Pr(|X[n]| < s)$, which is close to 0, the final outcome will be s.

To estimate the parameters p and q in dark chess, we performed a self-play experiment on FLIPPER vs. FLIPPER. There were 100 games played, and the results were 21 wins, 20 losses, and 59 draws. It is reasonable to set $x = 1$, $p = 0.2$ and $q = 0.6$ in dark chess. Thus, $1 - p - q = 0.2$ and $\sigma_n = \sqrt{0.8n}$.

In the experiments, the standard deviation σ_{100} is 8.9. The results of the first experiment are shown in Table 2. The performance of FLIPPER_att is significantly better than that of the original version of FLIPPER.

The results of the second experiment, shown in Table 3, suggest that revealing central pieces and increasing their mobility are critical policies in the opening book for the first player. Because the first player can reveal one more piece, it is important to strengthen that piece's control over other pieces with the above two policies. The second player needs to gain equal power by making a sequence of attacks. Understanding possible defensive moves is very important. When a player's attack fails, he needs to find new territory so that his power is approximately the same as that of the first player. Thus, we construct asymmetrical opening books, FLIPPER_F and FLIPPER_S, for first player and second player, respectively. The resulting opening book for FLIPPER_F has $50, 410, 385$ nodes and that for FLIPPER_S has $62, 908, 809$ nodes.

The results of the third experiment (see Table 4) show that the non-attack policies improve the performance of the opening books. Because the opening game is important to the outcome of dark chess, good strategies in the opening game will increase the probability that the player will win the whole game. After considering all the revealed places, the book suggests that $b7$ is the best place to reveal. Because $b7$ is a central place, it is a good location for controlling pieces in the corner. Note that if the opponent tries to launch a cannon attack, only two places can be used to attack $b7$.

Finally we would like to make the following four observations about the initiative in dark chess. (1) The first player can always reveal at least one of his pieces in a preferred location, but (2) the second player may decide to attack the piece immediately. In this scenario, (3) the first player has a slight advantage in terms of material; however, (4) the second player has an advantage because he can launch an immediate attack if he wishes to do so. The different advantages

Table 2. The results of 200 games played by FLIPPER_att and FLIPPER

Program	Win	Loss	Draw	s	$Pr(|X[n]| < s)$
FLIPPER_att first	29	15	56	14	0.934
FLIPPER_att second	25	13	62	12	0.901

Table 3. FLIPPER_att versus advanced versions using non-attack policies: F=the first player's opening book, S=the second player's opening book, X=not used

Program	Win	Loss	Draw	s	$Pr(|X[n]| < s)$	Selection
Flipper_cen first	26	20	54	6	0.731	F
Flipper_cen second	15	21	64	−6	0.731	X
Flipper_def first	19	28	53	−9	0.829	X
Flipper_def second	24	17	59	7	0.766	S
Flipper_mob first	22	18	60	4	0.652	F
Flipper_mob second	21	19	60	2	0.567	X
Flipper_ter first	16	21	63	−5	0.693	X
Flipper_ter second	32	19	49	13	0.919	S

Table 4. FLIPPER_F as the first player versus FLIPPER_att; FLIPPER_S as the second player versus FLIPPER_att.

Program	Win	Loss	Draw	s	$Pr(|X[n]| < s)$
FLIPPER_F vs. FLIPPER_att	25	18	57	7	0.766
FLIPPER_S vs. FLIPPER_att	24	17	59	7	0.766

of the players result in diverse policy selections. Exploiting the two advantages well is an important issue in the opening game.

6   Concluding Remarks

It is difficult to use a pure search algorithm to obtain a strong move in an opening game because of its large search space and probabilistic behavior. Automatically generated opening books are constructed off-line and can provide more precise information about opening positions and feasible strategies to use in the game. To the best of our knowledge, there are no published results or books for the opening game in dark chess. Using a brute-force method to construct an opening book is not practical due to time and space considerations.

In this paper, we propose a policy-oriented search method for compiling an opening book that contains useful information about opening strategies in dark chess. Since there is no opening game knowledge for humans or computers, an automatically compiled opening book can provide some guidelines about the importance of the opening game and how to play it well. Analyzing the generated opening book can also provide a clue about the advantage of the initiative.

The success of the policy-oriented method implies that the effective branching factor is significantly less than the number of valid moves in dark chess. By exploiting this property, it is possible to construct opening books automatically without sufficient prior knowledge. The method can also be extended to other games that have a large search space in the opening game.

References

1. Chen, B.N., Shen, B.-J., Hsu, T.-S.: Chinese dark chess. ICGA J. **33**(2), 93–106 (2010)
2. Lincke, T.R.: Strategies for the automatic construction of opening books. In: Marsland, T., Frank, I. (eds.) CG 2001. LNCS, vol. 2063, pp. 74–86. Springer, Heidelberg (2002)
3. Buro, M.: Toward opening book learning. ICGA J. **22**(2), 98–102 (1999)
4. Chaslot, G.M.J-B., Hoock, J-B., Perez, J., Rimmel, A., Teytaud, O., Winands, M.H.M.: Meta monte-carlo tree search for automatic opening book generation. In: Proceedings of the IJCAI09 Workshop on General Intelligence in Game Playing Agents, pp. 7–12 (2009)
5. Lincke, T.R.: Position-value representation in opening books. In: Schaeffer, J., Müller, M., Björnsson, Y. (eds.) CG 2002. LNCS, vol. 2883, pp. 249–263. Springer, Heidelberg (2003)
6. Gaudel, R., Hoock, J.-B., Pérez, J., Sokolovska, N., Teytaud, O.: A principled method for exploiting opening books. In: van den Herik, H.J., Iida, H., Plaat, A. (eds.) CG 2010. LNCS, vol. 6515, pp. 136–144. Springer, Heidelberg (2011)
7. Audouard, P., Chaslot, G., Hoock, J.-B., Perez, J., Rimmel, A., Teytaud, O.: Grid coevolution for adaptive simulations: application to the building of opening books in the game of go. In: Giacobini, M., et al. (eds.) EvoWorkshops 2009. LNCS, vol. 5484, pp. 323–332. Springer, Heidelberg (2009)
8. Walczak, S.: Improving opening book performance through modeling of chess opponents. In: Proceedings of the 24th ACM Annual Computer Science Conference, pp. 53–57 (1996)
9. Donninger, Ch., Lorenz, U.: Innovative opening-book handling. In: van den Herik, H.J., Hsu, S.-C., Hsu, T., Donkers, H.H.L.M.J. (eds.) CG 2005. LNCS, vol. 4250, pp. 1–10. Springer, Heidelberg (2006)
10. Chen, J.C., Hsu, S.C.: Construction of online query system of opening database in computer CHINESE CHESS. In: The 11th Conference on Artificial Intelligence and Applications (2001)
11. Flipper: A CHINESE DARK CHESS (DARK CHESS) program (2010). http://www.grappa.univ-lille3.fr/icga/program.php?id=637
12. Hsu, T.S.: Slides for the course theory of computer games (2012). http://www.iis.sinica.edu.tw/~tshsu/tcg2012/slides/slide13.pdf

Optimal, Approximately Optimal, and Fair Play of the Fowl Play Card Game

Todd W. Neller[(✉)], Marcin Malec, Clifton G.M. Presser,
and Forrest Jacobs

Department of Computer Science, Gettysburg College,
Gettysburg, PA 17325, USA
{tneller,cpresser}@gettysburg.edu,
{khazum,forrestjacobs}@gmail.com
http://cs.gettysburg.edu/~tneller

Abstract. After introducing the jeopardy card game Fowl Play, we present equations for optimal two-player play, describe their solution with a variant of value iteration, and visualize the optimal play policy. Next, we discuss the approximation of optimal play and note that neural network learning can achieve a win rate within 1 % of optimal play yet with a 5-orders-of-magnitude reduction in memory requirements. Optimal komi (i.e., compensation points) are computed for the two-player games of Pig and Fowl Play. Finally, we make use of such komi computations in order to redesign Fowl Play for two-player fairness, creating the game Red Light.

1   Introduction

Designed by Robert Bushnell and published by Gamewright in 2002, Fowl Play[TM] is a jeopardy card game variant of the folk jeopardy dice game Pig [4–6]. The object of Fowl Play is to be the first of 2 or more players to reach a given goal score. We focus here on the two-player game where the goal score is 50 points. A shuffled 48-card deck contains cards of two types: 42 chickens and 6 wolves. Each turn, a player repeatedly draws cards until either a wolf is drawn or the player holds and scores the number of chickens drawn, i.e., the *turn total*. During a player's turn, after the first required draw[1], the player is faced with two choices: *draw* or *hold*. If the player draws a wolf, the player scores nothing and it becomes the opponent's turn. If the player draws a chicken, the player's turn total is incremented and the player's turn continues. If the player instead

---

[1] Although a turn's initial draw requirement is not clearly or explicitly stated, the rules seem to imply the requirement, and it is necessary to avoid stalemate. Although the rules state "You can stop counting and collect your points at any time as long as you don't turn over a wolf!", there is an implication that one has *started* counting chickens/points. Consider the scenario where players have scores tied at 49 and the deck contains a single wolf card. It is in neither player's interest to draw the wolf card, so rational players would infinitely hold as a first action if permitted.

H.J. van den Herik et al. (Eds.): CG 2013, LNCS 8427, pp. 233–243, 2014.
DOI: 10.1007/978-3-319-09165-5_20, © Springer International Publishing Switzerland 2014

chooses to hold, the turn total is added to the player's score and it becomes the opponent's turn. All cards drawn are discarded after a player's turn. If the last, sixth wolf is drawn, the entire deck is reshuffled before the next turn begins.

For such a simple card game, one might expect a straightforward optimal strategy as with blackjack (e.g., "Stand on 17," etc.). As we shall see, this simple card game yields a much more complex and intriguing optimal policy, computed and presented here for the first time.

We begin by presenting play that maximizes the expected score per turn, explaining why this differs from optimal play (Sect. 2). After describing the equations of optimal play and the technique used to solve them, we present visualizations of such play and compare it to optimal play of a simpler, related dice game Pig (Sect. 3). Next, we turn our attention to the use of function approximation in order to demonstrate the feasibility of a good, memory-efficient approximation of optimal play (Sect. 4). Finally, we apply our analytical techniques towards the parameterization, tuning, and improved redesign of the game with komi for optimal fairness (Sect. 5). In Sect. 6 we discuss an analogous game, i.e., the game of Red Light. We give an overview of the results and directions of future research in Sect. 7.

2   Maximizing Score

The game of Fowl Play is simple to describe, but is it simple to play well? More specifically, how can we play the game optimally? This, of course, depends on how we define "optimal". As a first step, let us suppose that a player wishes to maximize score gain per turn.

Let us begin by defining relevant variables to describe each game state:

- i - current player score
- j - opponent player score
- k - turn total
- w - wolves drawn since last shuffle
- c - chickens drawn since last shuffle

We assume that a player with a turn total sufficient to win will hold. Thus, for non-terminal states, $i + k < g$, where g is the goal score of 50. Also, one cannot have drawn more chickens on one's turn than have been drawn since the last deck shuffle, so $k \leq c$.

Further, let $w_{total} = 6$ and $c_{total} = 42$ denote the total number of wolf and chicken cards, respectively. Let $w_{rem} = w_{total} - w$ and $c_{rem} = c_{total} - c$ be the remaining number of wolf and chicken cards in the deck, respectively.

In order to maximize the score, each decision to draw should be expected, on average, to increase the turn score. The expected gain should exceed the expected loss. The expected gain from a draw is the probability of drawing a chicken times the turn total gain of 1: $\frac{c_{rem}}{w_{rem}+c_{rem}}$. The expected loss from a draw is the probability of drawing a wolf times the turn total itself: $\frac{w_{rem}}{w_{rem}+c_{rem}}k$.

Thus, a player maximizing the score draws a card as required at the beginning of a turn when the expected gain exceeds the loss:

$$\frac{c_{\text{rem}}}{w_{\text{rem}} + c_{\text{rem}}} > \frac{w_{\text{rem}}}{w_{\text{rem}} + c_{\text{rem}}} k.$$

Simplifying, this condition is equivalent to:

$$c_{\text{rem}} > w_{\text{rem}} k.$$

So the player maximizing the expected score opts to draw a card when the number of chickens remaining exceeds the number of wolves remaining times the turn total.

However, there are many circumstances in which one should deviate from this score-maximizing policy. *Risking points is not the same as risking the probability of winning.* Put another way, playing to maximize points for a single turn is different from playing to win.

To illustrate this point, consider the following example. Assume your opponent has 49 points. There are 2 remaining cards in the deck: 1 wolf and 1 chicken. You have a score of 47 and a turn total of 2. Drawing would give an expected score gain of $\frac{1}{2}$ and a greater expected score loss of 1, so the score-maximizing player would hold.

However, the player playing to maximize the probability of winning will draw a card in this situation. If a chicken is drawn with probability $\frac{1}{2}$, the player then holds and wins, so the probability of winning with a draw is at least $\frac{1}{2}$. In fact, the probability is greater than $\frac{1}{2}$, as a draw of a wolf would result in a reshuffle and a non-zero probability that the opponent will draw a wolf on their next turn, allowing for a possibility of the player winning in some future turn.

Compare this with a decision to hold with 2 points instead. The opponent begins the turn with a tied 49-49 score and the same deck situation. By the same reasoning, the opponent has a greater than $\frac{1}{2}$ probability of winning, so in choosing to hold, the player would be choosing a probability of winning strictly less than $\frac{1}{2}$ and thus strictly less than the probability of winning by choosing to draw.

In the next section, we form the equations that describe optimal play, and show how it significantly deviates from and outperforms score-maximization play.

3   Maximizing the Probability of Winning

Let $P_{i,j,k,w,c}$ be the player's probability of winning if the player's score is i, the opponent's score is j, the player's turn total is k, and the wolf and chicken cards drawn since the last shuffle are w and c, respectively. In the case where $i + k \geq 50$, $P_{i,j,k,w,c} = 1$ because the player can simply hold and win. In the general case where $0 \leq i,j < 50$ and $k < 50 - i$, the probability of an optimal player winning is

$$P_{i,j,k,w,c} = \begin{cases} P_{i,j,k,w,c,\text{draw}} & k = 0, \text{and} \\ \max\left(P_{i,j,k,w,c,\text{draw}}, P_{i,j,k,w,c,\text{hold}}\right) & \text{otherwise,} \end{cases}$$

where $P_{i,j,k,w,c,\text{draw}}$ and $P_{i,j,k,w,c,\text{hold}}$ are the probabilities of winning if one draws and holds, respectively. These probabilities are given by:

$$P_{i,j,k,w,c,\text{draw}} = \frac{c_{\text{rem}}}{w_{\text{rem}}+c_{\text{rem}}} P_{i,j,k+1,w,c+1}$$
$$+ \frac{w_{\text{rem}}}{w_{\text{rem}}+c_{\text{rem}}} \begin{cases} 1 - P_{j,i,0,w+1,c} & w < w_{\text{total}} - 1, and \\ 1 - P_{j,i,0,0,0} & \text{otherwise.} \end{cases}$$

$$P_{i,j,k,w,c,\text{hold}} = 1 - P_{j,i+k,0,w,c}$$

The probability of winning after drawing a wolf or holding is the probability that the other player will not win beginning with the next turn. The probability of winning after drawing a chicken depends on the probability of winning with incremented turn total and chickens drawn.

At this point, we can see what needs to be done to compute the optimal policy for play. If we can solve the equations for all probabilities of winning in all possible game states, we need only compare $P_{i,j,k,w,c,\text{draw}}$ with $P_{i,j,k,w,c,\text{hold}}$ for our current state and draw or hold depending on which gives us a higher probability of winning.

Solving for the probability of a win in all states is not trivial, as dependencies between variables are cyclic. Beginning with a full deck, two players that manage to draw a wolf at the end of the next 6 turns will find themselves back in the exact same game state. Put another way, game states can repeat, so we cannot simply evaluate state win probabilities from the end of the game backwards to the beginning.

3.1   Solving with Value Iteration

Value iteration [1,2,8] is an algorithm that iteratively improves estimates of the value of being in each state until such estimate updates reach a terminal condition. We can summarize the value iteration algorithm of [8] by saying that state values are computed from Bellman's optimality equations by iteratively computing the right-hand-side equation expressions using the previous (initially arbitrary) value estimates, and assigning these new estimates to the left-hand-side equation state value variables. With each iteration, one tracks the maximum magnitude of change to any state value estimate. When the largest estimate change in an iteration falls below a desired threshold, the estimates have sufficiently converged and the algorithm halts.

In our application of value iteration as in [4], there are only two points to be made. The first is that each state value is defined to be the probability of winning with optimal play from that state forward. Winning and losing terminal states thus have state values of 1 and 0, respectively. The second point is that we generalize Bellman's equations to model a two-player non-drawing game as described in our previous optimality equations.

A similar technique is employed to evaluate relative player strengths. Given a fixed play policy for two players, we form equations to compute each player's probability of winning in all possible non-terminal states as in [6].

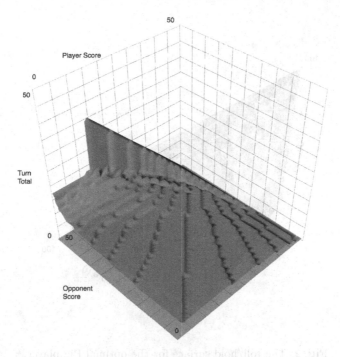

Fig. 1. The draw/hold surface for the optimal Fowl Play player when no cards have been drawn.

The score-maximizing player ("max-score player") makes the same decision as the optimal player in 90.04 % of the 10,487,700 non-terminal game states. As a result, the max-score player wins with probability 0.459 against the optimal player (probability 0.484 as first player and 0.434 as second player).

3.2 Visualizing the Solution

We can think of the optimal policy as being a boundary between draw and hold conditions in the 5-dimensional state space, yet this is not easily visualized. What we do to give the reader the best sense of optimal play is to present a selection of 3-dimensional graphs where 2 of the 5 dimensions, w and c, have been fixed. See Fig. 1, where $w = c = 0$ at the beginning of the turn. This is the situation both at the beginning of the game and when the deck has been reshuffled after the 6th wolf has been drawn.

For the given initial values of w and c at the beginning of the turn, the vertical height of the surface represents the number of cards a player should draw before holding for every possible pair of player and opponent scores.

As one varies the number of chickens and wolves drawn before the beginning of the turn, one can see that a player's play varies considerably and especially at score extremes (see Fig. 3).

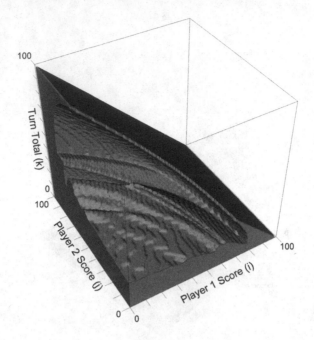

Fig. 2. The roll/hold surface for the optimal Pig player.

Fowl Play may be thought of as a card game variation of the folk jeopardy dice game Pig [4]. In the game of Pig, the first player to reach 100 points wins. Each turn, a player rolls a 6-sided die until either (1) the player holds and scores the sum of the rolls, or (2) rolls a 1 ("a pig") and scores nothing for the turn.

Fowl Play has many similarities to Pig. Both games have a similar first-player advantage in 2-player optimal play with the probability of the first of two optimal players winning being 53.06 % and 52.42 % for Pig and Fowl Play, respectively. Both games have a similar number of expected actions for an optimal two-player game with 167.29 and 164.98 expected player actions for Pig and Fowl Play, respectively. One can see some similarity in general hold boundaries, with the player ahead/behind taking lesser/greater risks.

However, Fowl Play has important differences as well. In Pig, there is no limit to the potential score of a turn (see Fig. 2), whereas in Fowl Play, the score is necessarily limited to the number of chicken cards remaining in the deck. The shared deck introduces interesting dynamics to play, as the outcome of one's turn determines the initial situation and scoring potentials for the opponent's next turn.

Of particular interest are situations where it becomes rational to draw a wolf after all chickens have been drawn. There exist situations where, with few chickens and an even number of wolves remaining, where a player drawing the last chicken should deliberately draw one of the remaining wolves, forcing a chain

Wolves Drawn

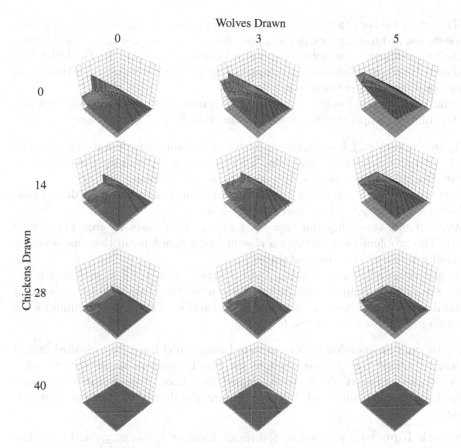

Fig. 3. The draw/hold surface with varied numbers of wolves and chickens drawn from the deck. The axis of the graphs are defined as the previous figure.

of wolf draws that leaves the player with a fresh deck and a greater upside for scoring.

Whereas simple mental arithmetic suffices closely to approximate optimal play for Pig, we have yet to find sufficient features that would allow mental arithmetic closely to approximate optimal play for Fowl Play.

4   Approximating Optimal Play

Having computed an optimal player for two-player Fowl Play, we observe that the computation requires the storage of 10,487,700 floating point numbers. The policy need not be represented as 10,487,700 boolean draw/hold values, however. We may simply map each (i, j, w, c) tuple to the *hold value*, the minimum turn total k for which the player holds. This reduces space requirements for storing the optimal policy to $50 \times 50 \times 6 \times 43 = 645,000$ integers.

However, this is still a large memory requirement for modern mobile and embedded systems. As an assignment for an upper-level elective course on advanced game artificial intelligence, students were provided the optimal policy for Fowl Play and challenged to create a function approximation that achieves near-optimal play with a very small memory requirement.

The reader should note that there are a number of possible design choices for the function approximation task. We describe four of them below.

- Approximate $V(i, j, k, w, c)$, the state value function equivalent to the expected win probability with optimal play. Then using the optimality equations, one can determine which next action maximizes the function.
- Approximate $Q(i, j, k, w, c, a)$, the action value function, where $a \in \{draw, hold\}$. Note that $\max_a Q(i, j, k, w, c, a) = V(i, j, k, w, c)$.
- Approximate the policy function $\pi(i, j, k, w, c)$ that maps to *draw* (1) or *hold* (0). The problem then becomes a classification problem, in that one seeks to classify states as *draw* or *hold* states.
- Approximate a function $draws(i, j, w, c)$ derived from $\pi(i, j, k, w, c)$ that maps the state at the beginning of the turn (when the turn total $k = 0$) to the total number of *draw* actions taken before a *hold* action. This is the minimum k for which $\pi(i, j, k, w, c + k)$ maps to *hold*.

Two students independently demonstrated that multi-layer feed-forward neural networks showed the greatest promise for closely approximating optimal play with minimal memory requirements. Since that time, the approach has been refined and further improved. We here present the details of the best approach to date.

Network Input: In addition to the input features i, j, k, w_{rem}, and c_{rem}, the features $\frac{c_{\text{rem}}}{w_{\text{rem}} + c_{\text{rem}}}$ and $\frac{w_{\text{rem}}}{w_{\text{rem}} + c_{\text{rem}}} \times k$, the probability of drawing chicken and wolf cards, respectively, aided in the function approximation. These seven features scaled to range $[-1, 1]$ form the input to the network.

Network Output: The single output unit classifies *draw* or *hold* actions according to whether the output is above or below 0.5, respectively. For training purposes, target outputs are 1 and 0 for *draw* and *hold*, respectively.

Network Architecture: Optimizing the number of hidden units empirically so as to boost learning rate and generalization, the single hidden layer worked best with 13 units. Thus, with an additional input and hidden layer bias unit, there are $8 \times 13 = 104$ weights from the input layer to the hidden layer, and 14 weights from the hidden layer to the single output unit, for a total of 118 weights. The activation function for all hidden and output units is the logistic function $f(x) = \frac{1}{1+e^{-x}}$.

Network Training: Network weights are initialized using the Nguyen-Widrow algorithm [7]. State inputs are generated through simulations of games with optimal play so as to focus learning on the most relevant parts of the state space. For each non-terminal decision state input, the network output is computed.

Optimal play is then referenced, and error is computed as the difference between 1 (*draw*) or 0 (*hold*) and the network output. Standard backpropagation [3] is then applied with learning rate $\alpha = 0.1$ to update network weights.

Online learning occurs for a duration of 100,000 games. Performance is then evaluated through network play against an optimal player for 1,000,000 games. If the win rate is better than 49.5 % and the highest win rate so far, we re-evaluate using policy iteration with $\epsilon = 1 \times 10^{-6}$. We then continue to alternate between training and evaluation for total of 400 iterations.

The network resulting from such training achieves an average win rate of 49.58 ± 0.10 % against an optimal player over 180 re-runs with a highest win rate of 49.79 %. This represents an impressive 5-orders-of-magnitude compression of optimal play policy information with very little loss of play quality.

5   Optimal Komi

Given that computation of optimal play is feasible for Fowl Play, we now turn our attention towards further analysis of the game, and parameterization of the game in order to tune the game for fairness and length.

Pig and Fowl Play both have a significant first player advantage. In fact, we can compute the optimal komi[2], i.e., compensation points, for the second player such that the first player win probability most closely approximates $\frac{1}{2}$ between optimal players. At the same time, we can also compute the expected number of game actions as an indicator of the expected game length.

For example, the game of Pig played between two optimal players will result in a first player wins of 53.06 % of the games after 167.29 expected player actions. However, if the second player starts with a komi of 4 points, the first player wins 50.16 % of the games after 164.01 expected player actions.

With optimal Fowl Play, the first player wins 52.42 % of the games after 164.98 expected player actions. However, if the second player starts with a komi of 1 point, the first player wins 50.54 % of games after 163.23 expected player actions.

Just as the commercial card game Fowl Play derives from the folk dice game Pig, we next apply our analytical methods to redesign Fowl Play as a new jeopardy game that we call Red Light.

6   The Game of Red Light

Red Light is a jeopardy game requiring red poker chips ("red lights"), green poker chips ("green lights")[3], and a bag or container that allows players to shuffle and randomly draw concealed chips.

---

[2] Komi is a Japanese Go term, short for "komidashi".

[3] For players with red/green color-blindness, we recommend use of yellow or light green chips for sufficient contrast.

For a two-player game, we suggest 4 red chips, 24 green chips, a goal score of 50, and a non-starting player initial score of 1.

The first player to reach the goal score wins. The starting player is determined by fair means (e.g., coin toss, alternation, etc.) A player begins a turn by drawing a poker chip from the bag or container.

If the drawn chip is a green light, the player sets it aside in a "turn total" pile. After drawing, the player then decides whether to draw again or hold. If the player holds, the number of green lights in the turn total pile are scored, the chips are placed in a discard pile, and play passes to the next player.

If the drawn chip is a red light, the player scores nothing for the turn, places all chips in the discard pile, and play passes to the next player. If the red chip was the last, fourth red light, discarded chips are returned to the bag or container and shuffled before the next turn.

The number of red and green chips for Red Light was chosen by varying the number of chips in play, analyzing the game, and optimizing komi. We found an extraordinarily good approximation of a fair game for 4 red chips and 28 chips total with a komi of 1, deviating from a fair win probability by 0.00001. With larger numbers of total chips, the game length shortens, and the ideal komi increases. We have opted for smaller numbers of chips (<50) in our redesign because poker chips are often sold in units of 25 or 50. The expected total number of player actions per game is 169.748, so Red Light has almost the same expected game length as Fowl Play.

Finally, we note that the two-player Red Light can effectively be played with standard playing cards in the same manner as Fowl Play using Ace through 7 of each suit. Aces can signify red lights, while all other cards signify green lights.

7 Summary and Future Research

In this paper, we have solved the jeopardy card game Fowl Play for the first time, demonstrating significant difference between score-maximizing and optimal play. We have visualized optimal play, highlighting similarities to and differences from its ancestor, the folk jeopardy dice game of Pig. Our work with neural networks has shown that it is possible to approximate optimal play well, reducing memory requirements by five orders of magnitude while only losing a fraction of a percent in win probability. Finally, we have parameterized the number of good and bad cards in the game as well as second-player komi in order to optimize the game for fairness. In this, we succeeded, creating an almost perfectly fair[4] variant of the game we call Red Light.

There are interesting questions that yet remain: How well does optimal 2-player play approximate optimal 3-player play if one treats the two opponents as a single abstract opponent? What human-playable strategies exist that significantly improve upon max-score play? How well can people play against an optimal player?

---

[4] i.e., fair after first-player determination.

We hope that other teachers and researchers find Fowl Play and/or Red Light of value in their work as well. They are excellent examples of games with minimal rules, yet complex optimal play.

References

1. Bellman, R.E.: Dynamic Programming. Princeton University Press, Princeton (1957)
2. Bertsekas, D.P.: Dynamic Programming: Deterministic and Stochastic Models. Prentice-Hall, Upper Saddle River (1987)
3. Bishop, C.M.: Neural Networks for Pattern Recognition. Oxford University Press, New York (1995)
4. Neller, T.W., Presser, C.G.M.: Optimal play of the dice game Pig. UMAP J. **25**(1), 25–47 (2004)
5. Neller, T.W., Presser, C.G.M.: Pigtail: a Pig addendum. UMAP J. **26**(4), 443–458 (2005)
6. Neller, T.W., Presser, C.G.M.: Practical play of the dice game Pig. UMAP J. **31**(1), 5–19 (2010)
7. Nguyen, D., Widrow, B.: Improving the learning speed of 2-layer neural networks by choosing initial values of the adaptive weights. In: International Joint Conference on Neural Networks, 1990 IJCNN , vol. 3, pp. 21–26, June 1990
8. Sutton, R.S., Barto, A.G.: Reinforcement Learning: An Introduction. MIT Press, Cambridge (1998)

Resource Entity Action: A Generalized Design Pattern for RTS Games

Mohamed Abbadi[1,2]([⊠]), Francesco Di Giacomo[1]([⊠]), Renzo Orsini[1]([⊠]),
Aske Plaat[2], Pieter Spronck[2], and Giuseppe Maggiore[2,3]

[1] Computer Science, Università Ca' Foscari DAIS, Venice, Italy
[2] Creative Computing, Tilburg University, Tilburg, The Netherlands
{mabbadi,fdigiacomo,orsini}@dais.unive.it
[3] Academy for Digital Entertainment, NHTV University of Applied Sciences,
Breda, The Netherlands
{a.plaat,p.spronck}@uvt.nl,
maggiore.g@nhtv.nl

Abstract. In many Real-Time Strategy (RTS) games, players develop an army in real time, then attempt to take out one or more opponents. Despite the existence of basic similarities among the many different RTS games, engines of these games are often built ad hoc, and code re-use among different titles is minimal. We created a design pattern called "Resource Entity Action" (REA) abstracting the basic interactions that entities have with each other in most RTS games. The paper discusses the REA pattern and its language abstraction. We also discuss the implementation in the Casanova game programming language. Our analysis shows that the pattern forms a solid basis for a playable RTS game, and that it achieves considerable gains in terms of lines of code and runtime efficiency. We conclude that the REA pattern is a suitable approach to the implementation of many RTS games.

1 Introduction

Real-time strategy (RTS) games have been highly popular for decades. As outlined by the ESA [10], RTS games are registering strong sales and a large number of play hours. Commercial RTS games are written by game developers of different backgrounds: from large studios to smaller independent developers of *indie games*. Indie developers [6] typically consist of small teams and their games are known for innovation [7], creativity [9] and artistic experimentation [2]. RTS games are also built as "serious games" [1], used for training and education, and as "research games" [3].

In general, the building of games is an expensive venture [4]. This is challenging in particular for indie developers and developers of serious and research games, who usually have only access to few resources. They would benefit of cost-effective development methodologies for games, through the identification and automation/reuse of common patterns in games. Surprisingly, from a survey of game development research and the corresponding literature, we noticed a

H.J. van den Herik et al. (Eds.): CG 2013, LNCS 8427, pp. 244–256, 2014.
DOI: 10.1007/978-3-319-09165-5_21, © Springer International Publishing Switzerland 2014

lack of studies of abstract patterns which characterize games, in particular RTS games. This motivated our research question: *To what extent can we capture the commonalities of RTS games in a re-usable design pattern?*

Section 2 discusses the essential elements of an RTS game. Section 3 specifies the *Resource Entity Action* (REA) design pattern [5] that captures these essential elements. Section 4 describes how the pattern is implemented as a language extension of the Casanova game programming language [8]. The language extension is purely declarative, using semantics that resemble SQL, providing an intuitive adoption. We implemented the extension in a full-fledged RTS, which we discuss in Sect. 5. Section 6 contains a summary and future work.

2   Essential Elements of RTS Games

RTS games are a variation of strategy games where two or more players aim to achieve specific (often conflicting) objectives by performing actions simultaneously in real time. The typical elements which arise from this genre are *units* (characters, armies), *buildings*, *resources*, and *battle statistics*. Players command units to perform different types of actions. These actions can affect several entities in the game world.

Units and buildings are the entities that players control to achieve their objectives. Units usually fight or harvest resources, while buildings may be used to create new units or research upgrades. Resources are gathered from the playing field and fuel the economy of the game entities. Battle statistics determine the offensive and defensive abilities of units in a fight. This taxonomy of the elements of an RTS game can be applied successfully to multiple games: Starcraft, C&C, and Age of Empires, as well as to all feature units, buildings, resources, and battle statistics, amongst other elements.

In order to arrive at our design pattern we will now apply a simplification. Battle statistics can be interpreted as *resources*, as for instance: "the life of a unit is the cost for killing it, payable in attack power." We can also merge units and buildings together into a new category called *entities*. This leads us to a simpler view of an RTS game as a game that is based on Resources, Entities, and Actions. We define them below.

1. Resources: numerical values in the battle and economic system of the game. In this group we find the *attack*, *defense*, and *life* patterns of entities. Resources also cover building materials and costs of production, deployment of units, development of new weapons, etc. (Resources are scalars.)
2. Entities: container for resources. They have physical properties and, as for the game logic, the difference among them is only the interactions. These interactions take place with resource exchanges through the actions. (Entities are vectors.)
3. Actions: resource flow among entities. Our model can be viewed as a directed weighted graph where the nodes are the entities, the weights are the amounts of exchanged resources, and the edges are the actions, that is, the elements which connect entities to one another. (Actions are transformation matrices.)

Next, we discuss how we model Resources, Entities, and Actions.

3 The REA Design Pattern

In this section we will define a model for an algebra to show that the REA (Resource Entity Action) model can be reduced to a problem of linear algebra (see Sect. 3.1). We then show how games that use this model can be further simplified by linguistic constructs (see Sect. 3.2).

3.1 Action Algebra

An action consists of a transfer of resources from a source entity to one or more target entities. We require that each entity has a resource vector, which contains the current amount of resources of the entity. The resource vector is sparse, since most actions involve only a few resource types. An action is expressed by a transformation matrix A.

Consider a set of target entities $T = \{t_1, t_2, ..., t_n\}$, which are the targets of the action, and a source entity e. Each entity t_i (including the source entity type) has a resource vector $\mathbf{r_i} = (r_{i_1}, r_{i_2}, ..., r_{i_m})$. The source entity also has a transformation matrix A of size $m \times m$, which defines the interactions between all the resources of the source entity and all the resources of the target entities. Moreover, we consider an integrator dt which contains the time difference between the current frame and the previous one. We then compute $\mathbf{w_e} = (w_{e_1}, w_{e_2}, ..., w_{e_m}) = \mathbf{r_s} \times A \cdot dt$. From the definition of matrix multiplication, it immediately follows that each component of $\mathbf{w_e}$ represents how a resource will change by applying the effect of all the other resources to it. We compute the vector $\mathbf{r_i'} = \mathbf{r_i} + \mathbf{w_e} \ \forall e_i \in E$ which replaces the resource vector in each target entity.

For instance, consider the action of a spaceship entity using laser to damage (resource) an enemy spaceship (entity). This involves a vector resource of two elements: laser and life points. The action must transfer laser points to subtract from the enemy life points. Suppose that the vector resource of the targeting ship is $r_s = (20, 500)$ and the vector resource of the targeted ship is $r_t = (15, 1000)$. Let the transformation matrix be $A = \begin{bmatrix} 0 & -1 \\ 0 & 0 \end{bmatrix}$ which means that the source entity will affect the life of the target with a negative number of laser points. Thus $w_e = r_s \times A \cdot dt = (20, 500) \times A \cdot dt = (0, -20) \cdot dt$. At this point, assuming $dt = 1$ second, we have $r_t' = r_t + w_e = (20, 1000) + (0, -20) \cdot dt = (20, 980)$.

3.2 A Declarative Language Extension

We now describe a language extension that implements the REA design pattern and its associated algebra for the Casanova game programming language [8]. The language extension is purely declarative. Its semantics are described using the SQL query language, which has the advantage of familiarity to most programmers.

Implementing the action algebra is done using an abstract class which contains an abstract method that performs the action. Each action is a class that extends the previous abstract class and implements the abstract method. This method will fetch the world looking for the information needed to find what entities are affected by the action execution. Each entity of the game will have a collection of actions that it can perform and automatically run by Casanova.

To identify the set of target entities T given a source entity and its action, we create a new type definition called *action*. An action is a declarative construct which is used to describe not only the resource exchange between entities, but also what kinds of entities participate in the exchange. The resource exchange is based on *transfers* (Add, Subtract, and Set), while the target determination is based on *predicates*: we filter the game world entities depending on their types, attributes and radius (specifying the distance beyond which the action is not applied). Some actions, called threshold actions, are not continuous and make use of special predicates to delay the execution (Output) until certain conditions are met.

Using actions it is possible to specify an exchange of resources in a fully declarative manner, so that the developer does not have to rewrite similar pieces of code ad hoc for each action.

4 Action Syntax and Semantics

Below we describe the syntax Sect. 4.1 and semantics Sect. 4.2 of actions in Casanova. The grammar allows the definition of actions, which make up the body of spacial Casanova entities that act as placeholders for actions. When an entity contains such an action, the Casanova runtime will apply it to all appropriate targets.

4.1 Action Grammar Definition

We first provide a taxonomy for actions. We divide the actions into three kinds: (1) constant transfer actions, (2) mutable transfer actions, and (3) threshold actions.

Constant Transfer actions update the target fields with a constant value or a value taken from one of the source fields. The source field is not affected by resource transfer. An example of a constant transfer action is a defense tower with infinite ammunition shooting an arrow at an infantry unit.

```
TARGET Infantry; RESTRICTION Owner <> Owner; RADIUS 1000.0;
    TRANSFER CONSTANT Life - ArrowDamage;
```

Mutable Transfer actions are used when the resource exchange transfers resources from the source entity to the target entity, or vice versa. An example of a mutable transfer action is a spaceship transferring minerals from its holds to a shipyard.

```
TARGET Shipyard; RESTRICTION Owner = Owner; RADIUS 150.0;
    TRANSFER MineralStash + Minerals;
```

Threshold actions follow the same transfer semantics as the previous two types of actions. In addition, there is more syntax information needed, since they have a collection of threshold values and output operations. The output operations are executed once when all the threshold values are reached. The threshold values are on fields belonging to the source entity. The output operations modify only fields of the source entity following the semantics of the transfer operations. An example for a threshold action is a worker building a town hall. When the `integrity` of the town hall reaches 100, a flag `completed` is set (which is one of its fields) which warns the system to replace the partially constructed building by the complete building.

```
TARGET ConstructionTownHall; RESTRICTION Owner = Owner;
    RADIUS 10.0; TRANSFER CONSTANT Integrity + 1.0;
    THRESHOLD Integrity = 100.0; OUTPUT Completed := true
```

Below we give a formal definition for the grammar instances presented in the examples above, using the extended Backus-Naur form. A *Casanova Entity* is an entity in the game world represented as a record; the special keyword `Self` is used to refer to the entity owning the action as one of its fields.

```
<Action> ::= TARGET <TARGET LIST> <RESTRICTION LIST> [<
    RADIUS CLAUSE>] <TRANSFER LIST>
    <INSERT LIST> [<THRESHOLD BLOCK>]
<TARGET LIST> ::= <ACTION ELEMENT>+
<ACTION ELEMENT> ::= Casanova Entity | Self
<RESTRICTION LIST> ::= {<RESTRICTION CLAUSE>}
<RESTRICTION CLAUSE> ::= RESTRICTION Boolean Expression of <
    SIMPLE PRED>
<SIMPLE PRED> ::= Self Casanova Entity Field (= | <>) Target
    Casanova Entity Field
<TRANSFER LIST> ::= {<TRANSFER CLAUSE>}
<TRANSFER CLAUSE> ::= (TRANSFER | TRANSFER CONSTANT)
(Target Casanova Entity Field) <Operator> ((Self Casanova
    Entity Field) | (Field Val)) [* Float Val]
<Operator> ::= + | - | :=
<RADIUS CLAUSE> ::= RADIUS (Float Val)
<INSERT LIST> ::= {<INSERT CLAUSE>}
<INSERT CLAUSE> ::= INSERT (Target Casanova Entity Field) ->
    (Self Casanova Entity Field List)
<THRESHOLD BLOCK> ::= <THRESHOLD CLAUSE>+
<OUTPUT CLAUSE>+
<THRESHOLD CLAUSE> ::= THRESHOLD
(Self Casanova Entity Field) Field Val
<OUTPUT CLAUSE> ::= OUTPUT
(Self Casanova Entity Field) <Operator> ((Self Casanova
    Entity Field) | (Field Val)) [* Float Val]
```

4.2   Formal Semantic Definition

Given the fact that actions resemble queries on entities, we specify their semantics as translation semantics to SQL. This allows us to leverage existing discussions on SQL correctness [12].

In defining our translation rules formally, we consider a set $T = \{t_1, t_2, ..., t_n\}$ of target types and a source entity type s. In all actions we select a subset of targets in each t_i on which to apply the action, using restriction conditions if any exist. After that we apply the resource transfer.

We assume that each entity type is represented by an SQL relation and that there exists a key attribute called **Id** for each relation. We now consider each of the three translation cases. In the translation rules we use notations inside the SQL code taken from the Backus-Naur form for grammar definitions. We also extend the SQL grammar with a global variable dt which is the time difference between the current and the last game frame. In this way the increment of the entity attribute values are proportional to the elapsed time. All types of actions evaluate the predicates in the restriction conditions and apply a filter to their targets. All targets further than the radius are automatically discarded when executing the action. The transfer predicates are executed immediately on all filtered targets.

For a **CONSTANT TRANSFER** we must update each target with the value in the source fields or constant values specified in the transfer clause. For simplicity, we assume that constant values are stored as attributes of the source entity.

Consider a set of resource attributes $A = \{a_{j_1}, a_{j_2}, ..., a_{j_m}\}$ of the source entity used to update the target t_i. To compute the contribution of all sources of the same type on the target t_i, we specify a relation of which the tuples represent the target id, followed by the total amount of resource a_{j_r} to transfer, called Σ_r:

Transfer
ID Σ_1 Σ_2 $\cdots$ Σ_m

The following SQL instruction implements the relation definition above.

```
SELECT    t_i.id , SUM(s.a_{j1}) AS Σ_1 ,
          SUM(s.a_{j2}) AS Σ_2 ,... ,
          SUM(s.a_{jm}) AS Σ_m

FROM      Target t_i , Source s
WHERE     <RESTRICTION LIST> [AND <RADIUS CLAUSE>]
GROUP BY  t_i.id
```

$\forall t_i \in T$ we update the target attributes $A' = \{a_{t_1}, a_{t_2}, ..., a_{t_m}\}$ using one of the target operators defined in the grammar (Set, Add, Subtract) with the attributes of the previous relation scheme.

```
WITH      Transfer AS(
               SELECT  t_i.id, SUM(s.a_{j1}) AS Σ_1,
                       SUM(s.a_{j2}) AS Σ_2,...,
                       SUM(s.a_{jm}) AS Σ_m)

FROM      Target t_i, Source s
WHERE     [<RESTRICTION LIST>] [AND <RADIUS CLAUSE>]
GROUP BY t_i.id)
UPDATE    Target t_i
SET       t_i.a_{t1} = u.Σ_1 | t_i.a_{t1} = t_i.a_{t1} + u.Σ_1 * dt | t_i.a_{t1} =
t_i.a_{t1} - u.Σ_1 * dt\
...

FROM      Transfer u
WHERE     u.id = t_i.id
```

For a **MUTABLE TRANSFER** the field of the source involved in the resource transfer is updated depending on the applied transfer operator. The resource is subtracted from the source field and added to the target field proportionally to dt, or vice versa.

To translate this semantic rule we must first determine how many targets (if any) are affected by each source entity, in order to obtain the following relation scheme.

TotalTargets
Source ID TargetCount

The SQL code implementing the previous scheme is the following.

```
TotalTargets =
SELECT  s.id,COUNT(*) AS TargetCount
FROM    Source s, Target t_1, Target t_2,...,Target t_n
WHERE   <RESTRICTION LIST> [AND <RADIUS CLAUSE>]
GROUP BY s.id
HAVING  COUNT(*) > 0
```

$\forall t_i \in T$ we need to obtain a relation storing what target each of the source entities is affecting, including a count of affected targets, using the following relation scheme.

OutputSharing
Source ID Target ID Output Sharing

This scheme is implemented by the following SQL code.

```
OutputSharing =
SELECT  *
FROM    TotalTargets c, SourceOutput c1
WHERE   c.s_id = c1.s_id
        AND SourceOutput =
            SELECT  s.id AS s_id,tᵢ.id AS t_id
            FROM    Source s, Target tᵢ
            WHERE   <RESTRICTION LIST> [AND <RADIUS
                    CLAUSE>]
        AND TotalTargets = [...]
```

Each target attribute receives an amount of resources equal to the total transferred resources divided by the number of targets. The complete SQL code to update the target t_i is the following.

```
WITH    Transfer AS(
        SELECT  tᵢ.id, SUM(s.aⱼ₁ / o.TargetCount) AS Σ₀,SUM(
            s.aⱼ₂ / o.TargetCount) AS Σ₂,...,SUM(s.aⱼₘ / o.
            TargetCount) AS Σₘ
        FROM    Source s, Target tᵢ,OutputSharing o
        WHERE   OutputSharing = [...] AND s.id = o.s_id AND
            t.id = o.t_id)
        GROUP BY tᵢ.id
UPDATE  Target tᵢ
SET     tᵢ.aₜ₁ = u.Σ₁ |tᵢ.aₜ₁ = tᵢ.aₜ₁ + u.Σ₁ * dt |tᵢ.aₜ₁ = tᵢ.aₜ₁ - u
    .Σ₁ * dt
...
FROM    Transfer u
WHERE   tᵢ.id = u.id
```

To update the Source relation we use a relation similar to the one use to update the target, but this time there is no need to save the count of the affected targets.

```
WITH    TotalTransfer AS(
        SELECT  s.id,s.aⱼ₁,s.aⱼ₂,...,s.aⱼₘ
        FROM    Source s, Target t₁,...,Target tₙ
        WHERE   <RESTRICTION LIST>
                [AND <RADIUS CLAUSE>]
        GROUP BY        s.id,s.aⱼ₁,s.aⱼ₂,...,s.aⱼₘ
        HAVING  COUNT(*) > 0)

UPDATE  Source s
SET     s.aⱼ₁ = s.aⱼ₁ - s.aⱼ₁ * dt|s.aⱼ₁ = s.aⱼ₁ + s.aⱼ₁ * dt
...
FROM    TotalTansfer u
WHERE   s.id = u.id
```

The **THRESHOLD** action is defined as the previous two types, i.e., it has a resource transfer definition which is always executed, and a set of threshold conditions that, if met, activate the Output operations, which are always towards the source entity. The attributes of the source entity affected by Output operations are updated with constant values, or with values from other attributes in the source entity. In the latter case the transfer is treated as for the mutable transfer case.

Consider a set of updating attributes $U = \{a_{k_1}, a_{k_2}, ..., a_{k_l}\}$ and a set of attributes to be updated $U' = \{a_{s_1}, a_{s_2}, ..., a_{s_l}\}$ in the output operation. We first check that all the conditions in the threshold clauses are met, then we update the attributes in the source entity appropriately.

```
WITH      TotalOutput AS(
          SELECT   s.id,s.a_{k_1},s.a_{k_2},...,s.a_{k_l}
          FROM     Source s
          WHERE    <THRESHOLD CLAUSE 1>
                   [AND <THRESHOLD CLAUSE 2>]
                     .
                     .
                     .
                   [AND <THRESHOLD CLAUSE l>])
UPDATE    Source s
SET       s.a_{s_1} = o.a_{k_1} | s.a_{s_1} = (s.a_{s_1} + o.a_{k_1})*dt; o.a_{k_1} = o.a_{k_1} -
          o.a_{k_1}*dt | s.a_{s_1} = (s.a_{s_1} - o.a_{k_1})*dt; o.a_{k_1} = o.a_{k_1} + o.
          a_{k_1}*dt
...
FROM      TotalOutput o
WHERE     s.id = o.id
```

5   Case Study

Below we present an RTS game that we used as a case study (see Sect. 5.1). It is created with Casanova, and the according benchmarks which test the action implementation. In the game, players must conquer a star system made up of various planets. Each planet builds fleets which are used to fight the fleets of the other players and to conquer more planets. A planet is conquered when a fleet of a player is near the planet and no other enemy fleet is defending it. An evaluation is given in Sect. 5.2.

5.1   Case Study

Three actions are required in this game. The first action, called **Fight Action**, defines how a fleet fights enemy fleets in range. The Fight Action subtracts $0.5 \cdot dt$ **life** points from the in-range enemy fleet during every frame (action tick).

```
Fleet = {Position: Rule Vector2;FightAction: FightAction;
     Owner: Ref Player;Life: Var float32;Fight: FightAction }
```

The Fight Action is defined as follows.

```
FightAction = TARGET Fleet; RESTRICTION Owner <> Owner;
    RADIUS 150.0; TRANSFER CONSTANT Life - 0.5;
```

The target is an entity of which the type is **Fleet**. The condition to execute the action is that the fleet must be an enemy (i.e., not the player). The attack range is 150 units of distance. 0.5 life points are subtracted for every attack. The second action is called BuildAction. It allows a planet to create a ship. In order to build a ship, a planet must gather 10 mineral units. Each planet has a field called GatherSpeed which determines how fast it gathers minerals. At every tick, the planet's mineral stash is increased by this amount. The action is a threshold action where the threshold value is a (fixed) number of the minerals of the planet. As soon as the threshold value is reached, we set the field NewFleet to TRUE (it is used by the engine to create a new fleet), and Minerals to 0 to reset the counter. The planet and its actions are as follows.

```
Planet = {Position: Vector2;Owner: Rule Ref Player;NewFleet
    : Rule bool;BuildAction:BuildAction;
    EnemyOrbitingFleetsAction : EnemyOrbitingFleetsAction;
    GatherSpeed: float32;Minerals: Var float32 }
```

```
BuildAction =
TARGET Self; TRANSFER CONSTANT Minerals + GatherSpeed;
    THRESHOLD Minerals 10.0; OUTPUT NewFleet := true; OUTPUT
    Minerals := 0.0
```

A Casanova rule is appointed to read the value of NewFleet and, if it is true, it spawns a new fleet.

The third action is required to check if a planet can be conquered by a fleet. A fleet can conquer a planet if there is no enemy fleet near the planet and if the fleet is sufficiently close. Thus the action definition is as follows.

```
EnemyOrbitingFleetsAction =
TARGET Fleet; RESTRICTION Owner Not Eq Owner; RADIUS 25.0;
    INSERT Owner -> EnemyOrbitingFleets
```

The action will add an enemy fleet sufficiently close to change the owner of the planet.

5.2   Evaluation

We evaluated the performance of our approach by the case study, and two additional examples: (1) an asteroid shooter, and (2) an expanded version of the case study with more complex rules. All were implemented in Casanova. Table 1 shows a code length comparison between the REA implementation and the standard Casanova rules for all three (case study, asteroid shooter, extended case study).

We note that in games with basic dynamics the code saving is low, due to the fact that there are only a few repeated patterns. The advantage of using REA

Table 1. CS (case study), asteroid shooter and expanded CS code length

	Game entities	Rules	Actions	Total
CS with REA	41	71	19	131
CS without REA	40	90	0	130
Asteroid shooter with REA	33	33	6	72
Asteroid shooter without REA	34	44	0	78
Extended CS with REA	135	138	40	313
Extended CS without REA	135	328	0	463

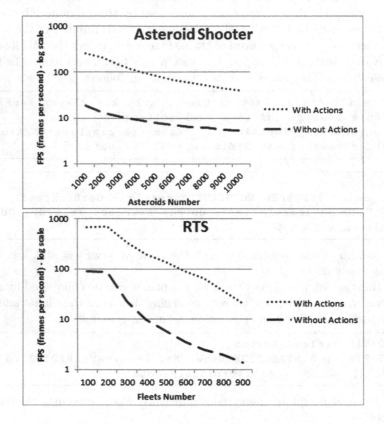

Fig. 1. Frame rate as a function of numbers of entities.

becomes evident in a game with actions involving many types of targets, such as the expanded case study. Furthermore, we managed drastically to increase the performance of the game logic. As Fig. 1 shows, using REA (labeled "with actions") results in a speedup factor of 6 to 25, due to the automated optimizations in the query evaluation. We also note that our implementation is flexible and general since it is possible to use actions to express a behavior, such as a projectile collision.

6   Summary and Future Work

In this paper we proposed the Resource Entity Action (REA) pattern to define RTS games. Use of this pattern should protect developers from writing and rewriting large amounts of boilerplate code. The paper presents the following four issues:

- the **REA design pattern** for making RTS games which reduces the interaction among entities to a dynamic exchange of resources;
- an expressive, declarative, high performance **language extension** to Casanova, with an appropriate grammar with new syntax and semantics resembling SQL;
- an evaluation of three examples which provides evidence for an increase in programming efficiency using REA; and
- an evaluation that shows an increase in run time efficiency of 6 to 25 times for the Casanova language, using a native code compiler/optimizer.

In future work, we estimate that even better results can be obtained with an actual access plan optimizer that increases the performance when exploring the structure of both the action query and the entity structure. Given the significant results on position indexing, the chance of defining multi-attribute indexes would increase the performance. Moreover, a system like F# quotations [11] may be used to increase the expressiveness of the actions.

References

1. Aldrich, C.: The Complete Guide to Simulations and Serious Games: How the Most Valuable Content Will Be Created in the Age Beyond Gutenberg to Google. Pfeiffer & Co., San Francisco (2009)
2. Andersen, E., Liu, Y.-E., Snider, R., Szeto, R., Popović, Z.: Placing a value on aesthetics in online casual games. In: Proceedings of the SIGCHI Conference on Human Factors in Computing Systems, CHI '11, pp. 1275–1278, New York, NY, USA. ACM (2011)
3. Buro, M.: Real-time strategy games: a new AI research challenge. In: Proceedings of the 18th International Joint Conference on Artificial Intelligence, pp. 1534–1535. Morgan Kaufmann Publishers Inc. (2003)
4. Buro, M., Furtak, T.: On the development of a free rts game engine. In: GameOn-NA 2005 Conference, pp. 23–27. Montreal (2005)
5. Gamma, E., Helm, R., Johnson, R., John, V.: Design Patterns: Elements of Reusable Object-Oriented Software. Addison-Wesley, Reading (1995)
6. Gril, J.: The state of indie gaming, vol. 4 (2008). http://www.gamasutra.com/view/feature/132041/the_state_of_indie_gaming.php?print=1
7. Kristiansson, J.: Interview starbreeze studios johan kristiansson, vol. 5 (2009). http://www.develop-online.net/features/478/Interview-Starbreeze-Studios-Johan-Kristiansson
8. Maggiore, G., Spanò, A., Orsini, R., Bugliesi, M., Abbadi, M., Steffinlongo, E.: A formal specification for casanova, a language for computer games. In: Proceedings of the 4th ACM SIGCHI Symposium on Engineering Interactive Computing Systems, EICS '12, pp. 287–292, New York, NY, USA. ACM (2012)

9. Michael, D.: Indie Game Development Survival Guide (Charles River Media Game Development). Charles River Media, Hingham (2003)
10. ESA Report: Essential Facts About the Computer and Video Game Industry (2011)
11. Syme, D.: Leveraging.NET meta-programming components from F#: integrated queries and interoperable heterogeneous execution. In: Proceedings of the 2006 Workshop on ML, pp. 43–54. ACM (2006)
12. von Bültzingsloewen, G.: Translating and optimizing sql queries having aggregates. In: Proceedings of 13th International Conference on VLDB, pp. 235–243 (1987)

Author Index

Abbadi, Mohamed   244
Althöfer, Ingo   84
Arneson, Broderick   60

Bonnet, Édouard   175
Buron, Cédric   187

Cazenave, Tristan   100, 187
Chang, Chieh-Min   1
Chang, Hung-Jui   151
Chen, Bo-Nian   221
Chikayama, Takashi   208

Di Giacomo, Francesco   244

Esser, Markus   125

Fernando, Sumudu   72

Graf, Tobias   14
Gras, Michael   125

Hayward, Ryan B.   60, 138
Horey, Therese   49
Hsu, Tsan-sheng   151, 221
Huang, Shih-Chieh   39, 60

Iida, Hiroyuki   199
Ikeda, Kokola   26, 199

Jacobs, Forrest   233
Jamain, Florian   175
Jouandeau, Nicolas   187

Kang, Hao-Hua   1

Lanctot, Marc   125
Lin, Hung-Hsuan   1
Lin, Ping-Hung   1

Lorentz, Richard   49
Lorenz, Ulf   110

Müller, Martin   39, 60, 72
Maggiore, Giuseppe   244
Malec, Marcin   233
Miwa, Makoto   208

Neller, Tod W.   233

Okaneya, Toshihisa   199
Opfer, Thomas   110
Orsini, Renzo   244

Pawlewicz, Jakub   60, 138
Plaat, Aske   244
Platzner, Marco   14
Presser, Clifton G.M.   233

Saffidine, Abdallah   100, 175, 187
Schadd, Maarten P.D.   125
Schaefers, Lars   14
Spoerer, Kristian Toby   199
Spronck, Pieter   244

Tanaka, Tetsuro   163
Tsuruoka, Yoshimasa   208
Turner, Wesley Michael   84

Ura, Akira   208

Viennot, Simon   26

Wei, Ting-Han   1
Winands, Mark H.M.   125
Wolf, Jan   110
Wu, I-Chen   1

Yamaguchi, Kazunori   163
Yamaguchi, Yoshiaki   163

Printed in the United States
By Bookmasters